ALLE ZEIT WACH
1842

Georges Matheron

Estimating and Choosing

An Essay on Probability in Practice

Translated by A. M. Hasofer

With 14 Figures

Springer-Verlag Berlin Heidelberg New York
London Paris Tokyo 1989

G. Matheron, Professor

Centre de Géostatistique et de Morphologie Mathématique
Ecole Nationale des Mines de Paris
35, rue St. Honoré
F-77305 Fontainebleau

A.M. Hasofer

School of Mathematics
Department of Statistics
The University of New South Wales
P.O. Box 1
Kensington, New South Wales, Australia

ISBN 3-540-50087-1 Springer-Verlag Berlin Heidelberg New York
ISBN 0-387-50087-1 Springer-Verlag New York Berlin Heidelberg

Library of Congress Cataloging in Publication Data.
Matheron, G. (Georges) [Estimer et choisir. English] Estimating and choosing/by G. Matheron; translated from the French by A. M. Hasofer. p. cm. Includes index.
ISBN 0-387-50087-1 (U.S.)
1. Probabilities. I. Title, QA273.M383813 1988, 519.2–dc19, 88-28159 CIP

Typesetting: Brühlsche Universitätsdruckerei,Giessen;
Printing: Saladruck, Steinkopf & Sohn, Berlin;
Bookbinding: Lüderitz & Bauer, Berlin
2161/3020-543210 – Printed on acid-free paper

Translator's Preface

Ever since the beginning of modern probability theory in the seventeenth century there has been a continuous debate over the meaning and area of applicability of the concept of probability.

The pioneers of the theory as applied to games of chance were able to sidestep the problem by introducing the concept of equally probable cases through symmetry, but it soon became apparent that this approach led to paradoxes when applied to continuous models.

The most popular interpretation of probability among statisticians has been for many years the frequency interpretation, which is quite satisfactory as long as the data represent outcomes of a large number of repeated trials. But since the middle of the century new areas of application of probabilistic models have appeared, for which a naive frequency interpretation is clearly deficient. This led to the development of the "subjective" interpretation of probability, which has become the predominant view among proponents of Bayesian Decision Theory.

My own interest in the subject started with my involvement with probabilistic design techniques and reliability calculations for large civil engineering structures. The interpretation of the concept of probability in that area presented grave difficulties for the following reasons:

(a) On the one hand the objects of study were *unique* complex structures, so that a frequency interpretation was quite unrealistic.

(b) On the other hand the essentially objective nature of engineering design made the adoption of any purely subjective interpretation repugnant.

It should be noted that exactly the same problem exists in all time series analysis work, since there also the data consists in just the one unique realization of a random function, and repeated trials are usually unavailable.

In view of the success of Bayesian methods in probabilistic design, a number of practitioners of the art did adopt the subjectivist viewpoint, albeit, in my opinion, rather reluctantly. There is in fact a widely held view that the only possible interpretation of a Bayesian procedure is a subjectivist one.

As for myself, I remained in a state of confusion about the matter until, in 1983, during a visit to the School of Mines in Fontainebleau, France, I came across a copy of Matheron's work "Estimer et choisir", which had been written in 1978 but had not had any impact on the English speaking statistical community, presumably because of language difficulties.

Reading Matheron's work was an illumination. Here was a coherent, well thought-out framework for the use of probabilistic models to describe unique phenomena in a purely objective way. In one sweep, Matheron was able to rid the practice of probability of all the confusing philosophical overtones that had clouded it for decades, and to provide a clear guide for the determination of the suitability and range of applicability of probabilistic modelling. Although "Estimer et choisir" had been written with Geostatistical applications in mind, the approach was of general applicability across the whole spectrum of probabilistic modelling.

I immediately wrote a paper entitled "Objective probabilities for unique objects", which appeared in "Risk, Structural Engineering and Human Error" (M. Gregoriu ed., University of Waterloo Press, 1984 pp. 1–17) and circulated it among my colleagues. But of course a 17 page article could not do justice to the breadth of material in Matheron's work.

What has spurred me to undertake the onerous task of translating the full text of Matheron's work into English was the appearance, at the beginning of 1986, in "Mathematical Geology" (Vol 18, No 1, 1986, pp. 94–117) of a virulent attack on probabilistic models in Geostatistics entitled "Matheronian Geostatistics – Quo Vadis?" by Philip and Watson. The paper questioned the very foundations of the rationality of using probabilistic models in all Engineering and perhaps all Science as well.

There were a number of responses to the paper that appeared in subsequent issues of Mathematical Geology, but it was felt that nothing short of an English translation of the full text of "Estimer et choisir" would be adequate to set matters right.

Matheron's masterpiece is, to my knowledge, the only full-fledged treatment of the foundations of practical probability modelling that has ever been written and thus fills an important gap in the literature of Probability and Statistics. Of course, some of the background material presented has been for some years part of the oral tradition current among experienced statisticians. But even that material is presented in a refresingly crisp and explicit manner.

I hope that practitioners of probabilistic modelling will read carefully Matheron's work, from which all can derive valuable insights into the foundations of the art. But most of all it should become required reading for all budding researchers who are using probabilistic models in their work.

A. M. Hasofer, March 1988

Contents

Part II Criteria of Internal Objectivity in the Case of Unique Phenomena

Part III Operational Reconstruction

Part I A Quest for Objectivity

"He who desires the absolute must accept in addition subjective and egocentricity, and he who wishes to remain objective cannot avoid the problem of relativism."
H. Weyl

Chapter 0

Introduction

This book is not a disinterested effort. Not only am I defending a thesis, but also a *practical methodology*, a collection of models, techniques and "tricks," which are sometimes of very tenuous orthodoxy. They have been developed [1] over the years to describe, study and estimate the most varied phenomena: cancer cells, forests; the structure of a rock or that of a metal alloy; an oil deposit or a road surface; pollution or meteorology; underwater cartography or geophysical prospecting. This methodology and its practice originate in a very down-to-earth problem: estimating mineral deposits. It is this particular area which has served as a testing ground, and has provided the most numerous and decisive verifications.

The problem of raw materials is apparently to-day at the heart of everyone's preoccupations. In practice, however, it soon appears that apart from a small number of mining industry specialists, very few people can cope with listening until the end to any serious presentation of the notion of recoverable reserves of a mineral deposit. It is even less feasible to go through the many really subtle difficulties which need to be elucidated one by one and then successfully overcome, before one can suggest a sensible estimation method for these reserves. The complexity of the new discipline which had to be hammered out to this end, and which has now proved itself, *Mining Geostatistics,* may well surprise any statistician or probabilist who would drum up the courage to acquaint himself with it. And the technicalities of the problem would probably discourage the best-intentioned reader. So I shall merely allude to them in what follows.

The phenomena in which we are interested have two important characteristics: firstly, they take place in physical space, the one in which we live. We say that they manifest themselves in a *regionalized* form. Secondly, most of the time we are dealing with unique objects or phenomena. Having handled and successfully solved a large number of particular cases, as confirmed by practice, we suspect that these objects are perhaps not as unique as they appear to be at the outset. The methodology which has enabled us to solve them, and which has now proved itself, undoubtedly possesses a certain form of objectivity, which we shall call *external objectivity* and which is simply the sanction of practice. But these objects remain unique as individuals which will never be found again elsewhere in exactly the same form. There do not exist two deposits, or two forests, or two stars, which are completely identical. To what extent does an estimation or a probabilistic model concerning *this* deposit or *that* forest (but not a deposit or a forest in general) possess an objective meaning? It is to a quest for the criteria of this *internal objectivity* that this essay is devoted.

I cannot therefore sidestep in any way the famous, but so badly formulated question: "Is probability subjective or objective?" In fact there is not, nor can there be, any such thing as probability in itself. *There are only probabilistic models.* In other words, randomness is in no way a uniquely defined, or even definable property of the phenomenon itself. It is only a characteristic of the model or models we choose to describe it, interpret it, and solve this or that problem we have raised about it. Depending on the nature of these problems, we may well adopt successively different models which, depending on the context, ascribe different probabilities to a given event whose definition remains apparently the same. The only real problem is to know whether a given model, within a given context, does or does not possess an objective meaining and to be able, if necessary, to perform a "sorting out" operation. By this I mean to be able to distinguish, among the various concepts, statements, parameters etc., which appear in the model, those which possess in what we call reality a counterpart which is objective, observable, measurable, etc. As for the others, which I shall call conventional (rather than subjective), they might play a useful heuristic part in the elaboration and working out of the model. But they will have to disappear from the final formulation of our conclusions and our results. For to the extent that the problems which we attempt to solve are real problems, the solutions that we propose must also respect the principle of reality, lest they turn out to be illusory. Let us note that there is no question of adopting a "conciliatory" position between the two extremes represented by the "objectivist" and "subjectivits" theses. We must cut into the quick. From the Bayesians I retain much: their fierce criticism of the implicit postulates of the objectivists, which in many cases cannot be said to be either true or false, but just devoid of meaning; also the fundamental role which they attribute, with good reason, to what they call "a priori information," that is, information which is nonnumerical, qualitative, of structural nature, and which tells us simply, *in short, what it is all about*. With one and the same sequence of zeros and ones, we choose different models depending on whether we are dealing with the results of a game of heads and tails, an alternation of rainy days and fine days, or the binary expression of a transcendental number. But I do not draw the conclusion that our probabilistic models must be abandoned without recourse to the arbitrariness of individual subjectivity. On the contrary I suggest that, as I said, a sorting out operation must be performed. Above all, we should *follow the lead of the physicists,* and strive towards on operational reconstruction of the concepts of our model, or at least of those which our preliminary investigation has identified as capable of conveying objective information.

This methodological position is reflected in the terminology which I am adopting. Orthodox statisticians would say: estimating and predicting. The Bayesians, that is, the subjectivists would say: evaluating and forecasting. I prefer: *estimating* and *choosing*. This deserves some comments. I use the verb "to estimate" whenever one must evaluate a magnitude which one does not know, but which nevertheless exists independently of ourselves and of the state of our information. Estimation, therefore, refers to an objective, physical magnitude. It does not matter whether it is associated in our model with a random variable or a "parameter." In fact, depending on circumstances, the same reality, for example the mean grade of a mineral deposit, will be treated in the model either as an objective parameter or as a random variable. But in either case I shall speak of estimating the mean grade (and not of predicting or forecasting it). I shall use the verb "to choose" sometimes in a neutral sense (without prejudging whether it will be appropriate later to attribute on objective or conventional meaning to the chosen parameter). At other times I shall use it in a limiting sense, when it will be necessary, for reasons of convenience, to attribute a numerical value to a parameter which is conventional and not objective. It will then be possible to choose that value in a more or less arbitrary manner without compromising the objectivity of the procedure as a whole. In short, we *estimate* objective magnitudes, we *choose* methods, models or conventional parameters and we *agree* on criteria. For the Bayesians, there is no estimation, but only "choices." They evaluate subjective probabilities and derive from them equally subjective forecasts concerning unknown events or magnitudes. Conversely, the orthodox, objectivist statisticians reserve the word "estimation" for the choice of the value of the parameters in their models, without ascertaining at the outset whether or not these parameters have an objective counterpart in reality. But they predict the future or unknown value of random variables, even when the latter represent actual physical reality. Their terminology conveys the impression of a kind of covert platonism: the model and its parameters are given to us from a Garden of Eden of ideas, and derive from this noble origin a substantial character of reality. Their ontological status is far superior to anything that can be claimed for their pale earthly imitations: the variables of the model, that is, the physical magnitudes by which reality is described. It is this implicit postulation of philosophical realism which the Bayesians are right to denounce. For us, it is only after a critical examination performed for each particular case that we shall bestow an objective status on this or that parameter, and that consequently we shall speak of its estimation.

The above remarks dictate the plan of this essay. In the first part, I examine the criteria and limitations of external (or methodological) objectivity, which is simply the long-run sanction of practice. I also examine internal objectivity, which is based, when all is said and done, on the concrete, physical characteristics of the individual object. At that stage, the examples I give will come from our common cultural background: rain and fine weather, a game of heads and tails, or protein molecules. From the second part onwards, I shall deal more specifically with phenomena which are regionalized in space, as well as with the "topo-probabilistic" models which represent them. Our aim will then be to isolate criteria for internal objectivity. How shall we choose a model, how shall we sort out what is objective from what is not? How shall we localize on this ascending curve the precise point at which we shall introduce, in practice, an *anticipatory hypothesis,* loaded with objective information which is not contained in the data? For it is this hypothesis which will allow us to solve the problems we are setting ourselves, even though it can never completeley guarantee us against a risk of *radical error*. These are the questions which I shall try to answer.

In the third part, I shall attempt, in some particular cases, to reconstruct in operational terms, after the manner of the physicists, some of the concepts which underlie our models. This reconstruction will allow us to specify in a more precise way the nature of this famous anticipatory hypothesis: it is merely a simple *approximation*, adopted as a hypothesis before its validity has been effectively checked. Hence the ever-present risk of radical error. Similarly, *"statistical inference"* will be stripped of much of its mystery; inasmuch as its target is a conventional parameter, it boils down to a false problem. But inasmuch as its target is a physical reality, it is merely an *approximate numerical evaluation* of a spatial integral.

The main casualty of this study will probably be the concept of conditional expectation. This is an exceedingly deep mathematical idea, which will however turn out to be almost incapable, in our context, of taking on the status of a physical concept. As a consolation prize, I shall propose a simpler substitute, namely the concept of a disjunctive estimator, and I shall also say a few words about the use of purely *heuristic models* which are acceptable and legitimate under certain conditions.

Although our probabilistic models enable us to tackle a variety of quantitative and qualitative [2] problems, I shall center my exposition mainly on estimation, for reasons of simplicity. The difficulties raised by the estimation problem are not of a mathematical nature. We encounter them at two extreme levels: at the initial stage of conceptualization, i.e. of model choice, and at the final stage of practical application.

At the initial stage we are confronted with inevitable questions concerning the meaning, objectivity and validity of all ulterior steps. Given a unique phenomenon which is completely determined even if largely unknown to us, is it at all possible, or meaningful, to conceptualize it as a probabilistic model, that is, to represent that phenomenon as drawn at random from a family of virtual phenomena deemed possible? Has the choice of that model an objective meaning, i.e. does it lend itself to one form or another of experimental control, or must it be abandoned, without recourse, to the arbitrariness of individual subjectivity? Finally, assuming that these experimental controls are possible, would they lead to positive results which would allow us to reach an unambiguous conclusion regarding the objective validity, full or partial, of the whole endeavour?

At the final stage, that is, at the level of practical applications, we are confronted with similar questions but in a more specific form. Starting with the information we dispose of to estimate, for example, a mineral deposit (information which is always fragmentary, and often of poor quality), can we specify a probabilistic model adequate for the given data? In particular, can we evaluate sensibly the numerical values of the parameters on which the model is based? Assuming that this is possible, to what extent can we then extrapolate from the known to the unknown i.e. draw valid conclusions about those parts of the deposit for which we have no information from the model which has been calibrated solely by the known data? It appears that we can only provide an answer to this decisive question (a very Kantian one), namely "How is estimation possible?" by accepting, in one form or another, a hypothesis of at least local statistical homogeneity. (One extreme form of this hypothesis, of exaggerated

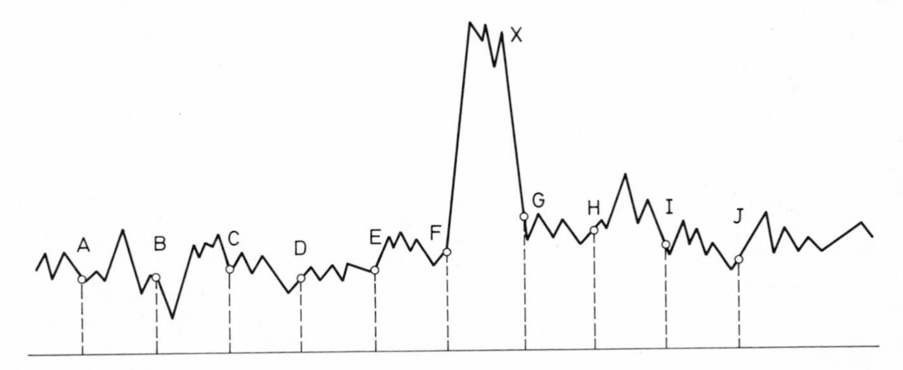

Fig. 1. A radical error is unavoidable

severity, is that of classical stationarity.) We expect the phenomenon to behave, where we do not know it, in a manner reasonably analogous to the available data we can observe. But even though this hypothesis is indispensable to the provision of a foundation for the possibility of estimation, it does not follow that it is always fulfilled. it is easy to provide counterexamples. In Fig. 1, where A, B, C, ...J represent the available data, one can visualise such a situation: nothing allows us to forecast the anomaly X, and there is here no possibility of estimation (by any method). A *radical error* is inevitable.

A few more words before concluding this rather long introduction. In what follows, there will not be found any "world view," whether explicit or implicit, but only methodological advice for practitioners of probability. One can, if necessary, distinguish between the *prime mover* and the *code*. The prime mover, i.e. the "dialectic" is always implicitly present. It incites us to progress, to start new endeavours, without ever remaining at rest. It enables us to "understand" all of the previous developments in a vast retrospective synthesis, but only with hindsight, like the Minerva bird which rises after sunset. It may be that this is just an anthropomorphic illusion. At any rate, this point of view is only suitable for a historian or a philosopher of science who attempts to reconstitute the stages covered by the human mind in its efforts to elaborate and reconstruct reality. Not so the practitioner: *in his work* he rather needs a code, a sort of plumb line to help him build straight walls. This does not mean that he subscribes to any particular philosophy of the plumb line, but only that he likes his job.

Notes

[1] Mainly at the Centre for Geostatistics and Mathematical Morphology of the School of Mines in Fontainebleau.

[2] See e.g. what we call Mathematical Morphology.

Chapter 1

Monod and the Concept of Chance or How to Overstep the Limits of Objectivity

In a serious work, written by a competent author, on a scientific subject, an expression such as "this phenomenon is due to chance" constitutes simply, in principle, an elliptic form of speech. It really means "everything occurs as if this phenomenon were due to chance," or, to be more precise: "To describe, or interpret or formalize this phenomenon, only probabilistic models have so far given good results." It is then only a statement of fact. Eventually, the author may add: "And it does not appear to me that this situation is likely to change in the foreseeable future." There is then a personal stand, a methodological choice, which opens certain possibilities to research, but closes others. One understands that the author, who knows his subject, since he has practised it for many years, wishes to spare his colleagues the trouble of entering a blind alley, thus saving them precious time.

There is a risk, because after all there is no reason to believe that tomorrow or in ten years time another researcher will not publish a deterministic theory which will explain beautifully and completely the phenomenon at hand, and we shall have perhaps missed the boat by following the implicit advice of our author: it is a real risk, but one which is normal and inherent to the practice of scientific work, since we must in any case make methodological choices.

In this purely operational sense, recourse to chance, that is, in realitiy, the decision to use probabilistic models, is perfectly legitimate and does not take us outside the framework of objectivity.

Sometimes, however, and probably more often in works of popularization and philosophical synthesis than in purely scientific works, we are presented with quite a different interpretation. It is suggested, nay even affirmed, that Chance (sometimes with a capital c) acts decisively in its own right on the course of events. This is no longer a methodological choice. Chance is now hypothesized, supplied with positive attributes, set up as a *deus ex machina*. Under these circumstances, to attribute the phenomenon to Chance, is equivalent to attributing it to Providence, and both are foreign to scientific methodology.

Chance is a Metaphysical Concept

Illegitimate use of scientific concepts beyond the limits within which they have an operative meaning is nothing else but a surreptitious passage into metaphysics. It is a danger that threatens even the greatest of us. Let us take a renowned work, written by an eminent biologist: "Chance and Necessity," by J. Monod. In the last page we find the following distressing statement: "The ancient alliance is broken; man finally knows that he is alone in the indifferent immensity of the Universe from which he has emerged by chance." Here the word "chance" is clearly used in a purely metaphysical (not operative) sense. Elsewhere [1], we can observe the very moment when the passage from one status to the other occurs:

> "We say that these alterations (i.e. genetic alterations) are accidental, that they occur *by chance* [2]. And since they constitute *the only* [3] possible source of modifications of the genetic text, which is in turn the *only* [3] depository of the hereditary structures of the organism, it necessarily follows that chance is the *only* [3] source of every novelty, of every creation in the biosphere. Pure chance, only chance, absolute but blind freedom is at the very root of the prodigious edifice of evolution: this central notion ... is the *only* [3] conceivable one, since it is the only one which is compatible with the facts of observation and experience."

The first sentence ("genetic alterations are due to chance") may perhaps still be understood in a scientific (i.e. operative) sense, and means in this case: experimental data in our possession are compatible with such and such a probabilistic model (which in any case should have been specified), and we do not have any deterministic model which can account for them in a more satisfactory fashion. However, starting from the second sentence, chance is credited with an active, positive role: it constitutes the only possible source of all novelty in the biosphere. In what sense can chance be said to be the "source" of anything? In order to see that there is here a surreptitious passage into metaphysics, it is only necessary to replace the word "chance" by its operational definition. We obtain approximately the following piece of nonsense: "the fact that, until now, only probabilistic models have turned out to be compatible with what we know of genetic alterations is the only source of all creation and all novelty in the biosphere." The uneasy feeling we are left with when we read this strange statement is due to the radical heterogeneity between the subject (a methodological finding, commensurate with our modest scientific practice and its limited scale) and the active and universal role which is attributed to it (the role of a cosmological Demiurge).

Monod was partly conscious of this dangerous confusion of genre. Immediately after the passage we have just quoted, he actually adds that "it is very important to state unequivocally the exact sense in which the word chance can and should be used when applied to mutations as the source of evolution." He then introduces a distinction between two types of uncertainty: an "operational uncertainty" and an "essential uncertainty." I am not sure that this is a relevant distinction. For, unless one is prepared to go outside the framework of objectivity, it would be necessary to define an operative criterion which would enable us to unequivocally discriminate between the two types of uncertainty. I doubt that this is possible. The proper empirical foundation on which we base our diagnosis of uncertainty is, in fact, the same in both cases: at the present stage of our knowledge, we are unable to predict with certainty such and such an event or such and such a phenomenon. In the first case we envisage that a considerable improvement in our knowledge or our research capabilities (which is at least theoretically conceivable, even if there is little hope of seeing it occur in practice), might enable us to remove the uncertainty. In the second case, we categorically deny this possibility. But it is we who have decreed this, not the empirical data we possess. They remain perfectly silent on that point. It is not a statement of fact, but a tacked-on interpretation, an extrapolation to infinity. And this in itself is illegitimate, even if it is based on serious theoretical considerations, because within the domain of empirical disciplines, one cannot extrapolate a theory to infinity, however strongly corroborated it might be, without ipso facto overstepping the boundaries within which the theory has an operational meaning and has received the sanction of experiment.

The Parable of the Plumber and the Physician

It is at any rate a remarkable fact that our eminent biologist has in no way thought it necessary to define an operational criterion which would enable us to distinguish between the two types of uncertainty. "The best thing to do," he says, "is to take some examples." The first example, a classical one, is that of a game of dice or roulette. It is quite clear that if we knew with sufficient precision the initial conditions (position, velocity and momentum of the die at the moment it leaves the dice-box) and the boundary conditions (the precise topography of the room where the game is being played), the laws of mechanics would allow us, in principle, to predict with complete certainty the number which would appear. However, it turns out that very small changes in these initial conditions, ones which are much smaller than any we would be able to control experimentally, are enough to modify the final result (e.g. from a 5 to a 6, etc.). We encounter here the condition of "inseparability of initial conditions" which characterizes, according to J. Ullmo[4] the introduction of probabilistic models in physics. It is undoubtedly an "operational, not an essential, uncertainty," and it is "purely for methodological reasons" that we have recourse to the calculus of probability. All this is indisputable, and in fact quite trivial. We look forward with some impatience to the second example, the one which will enlighten us about the rather mysterious nature of the second type of uncertainty, the "essential uncertainty." Unfortunately what follows is nothing more than the all-too-well known parable of the plumber Dubois who, while working on a roof, lets his hammer fall inadvertently. The latter cracks the skull of Doctor Dupont who happened to pass by on his way to an urgent call by one of his patients. No reader of "Chance and Necessity," this great and beautiful work, is able, I think, to avoid a feeling of uneasiness in finding such a weak argument coming from such an illustrious pen. It is this "parable" which is supposed to define the notion of "absolute coincidence," or, what amounts to the same thing, "essential chance," that is, according to Monod, who borrows there an idea which goes back to Cournot, "the total inherent independence of two sequences of events which, when they intersect, cause the accident."

It is the terminology itself which gives away the limitless extrapolation which occurs here. The coincidence is *absolute,* the independence is *total.* Have these words any meaning, when applied to approximate concepts, of limited range, such as coincidence or independence? There is moreover a certain inherent contradiction in the idea of two sequences of "totally" independent events, which nevertheless end up intersecting. The concept of "independence" used here is not even defined. We are only told that the two sequences of events have "nothing to do" with each other. But whatever the meaning of this vague phrase, it is clear that it is only from our viewpoint that these two sequences have "nothing to do with each other." It only shows that we are incapable of predicting the "intersection" and the "accident" which resulted from it, nothing more. In what sense does this unpredictability differ from that which governs a game of dice? There is a difference, but it belongs to the level of abstraction at which we position ourselves to observe the two types of events, and not to their nature. In the first case, we think of a game of dice in general, but in the second, we think of a unique, singular event, completely specified by the identity of the plumber and the physician. (One could have also added the place and date.) It is only the asymmetry of these two viewpoints which creates an impression of contrast. But it is possible to particularize as well the first viewpoint, the one concerning the game of dice. We can suggest the following "parable," which would be symmetrical to that of the plumber and the physician: "On such and such a day, at such and such an hour, Mr. Duval enters cafe X and meets his old friend Mr. Durand whom he has not seen for three years. They drink liberally to the happy occasion and then, at the suggestion of the barman, they engage in a lengthy game of dice. At the 26th throw, one die, which had been inexpertly thrown, ends up in the coffee cup which Mr. Dupont had just emptied and put on the counter, bringing up the number 6." We have here an event just as unpredictable, a "chance" occurrence just as "essential" as in the fable of the plumber and the physician. The moral of this trivial story is simply that any singular event which occurs in the world where we live, whenever sufficiently individualized by specifying space and time parameters, appears as the result of the intersection of not only two, but a very large number of "sequences of events", which have very little to do with each other. They derive from a "chance occurrence" which is neither more nor less essential than that which is invoked by J. Monod. And this simply amounts to the trivial statement that a unique and singular event does not belong, by its very nature, to the realm of scientific knowledge, because the requirement of "repeatability," which alone assures the objectivity of our empirical disciplines, is by defintion not fulfilled.

Conversely, if we mentally divest the unique and singular event of the anecdotal cicumstances which accompanied it (and in particular of the time and space parameters) so as to reduce it to a particular occurrence of a more general class of events capable of repeating themselves, we then by that very process *make* it into a possible object of scientific knowledge. We can then propose deterministic or probabilistic models and theories which are corroborated or refuted, depending on whether they will accord not only with the particular event, but also with all those which belong to the class which we have constructed, and which we have had the opportunity of observing. For example, we could be led to construct a probabilistic model containing at least three events: A would be the fall of a plumber's hammer at a given point in time and space. B would be the presence of the head of a passer-by at the same point in time and space. AB would be the intersection of the two preceding events. The assertion that the first two events are independent simply means that, *within this model,* the probability of AB is equal to the product of the probabilities of A and B i.e. that the well-known equation $P(AB)=P(A)P(B)$ holds. At this stage we are only dealing with a feature of our model and not with a statement of fact. To cross over and reach a statement which has an objective meaning, we must include in our model further features (e.g. time-wise stationarity). Through them we shall be able to identify legitimately the probability $P(A)$ of an event A with the empirical relative frequency $v(A)$ of its occurrence in time to a degree of approximation which the techniques of Mathematical Statistics enable us to evaluate in principle. It is only then that we shall have an objective statement, that is, one which can be experimentally verified, namely that the relative frequency $v(AB)$ of the product event must be equal, to the degree of approximation specified above, to the product $v(A)v(B)$ of the relative frequencies of the two events in question.

Incidentally, while we do not have the necessary statistical data, we may well foresee that in all likelihood, the equation $v(AB)=v(A)v(B)$ will not be verified. This is for the simple reason that when there are, in general, no plumbers on roofs (e.g. at night, on very cold days and on days of heavy rain etc...) there are also very few passers-by in the streets. Thus the two events which were presented to us as a typical example of "total independence" may well have, on analysis, a positive, statistically significant, correlation.

Polypeptide Chains

It is with a feeling of embarrassment that I put forward such heavy and at the same time trivial criticisms against a book I otherwise admire. But these criticisms are perhaps not completely useless: for if a scientific mind as great as that of J. Monod can let itself be caught in such a crude *epistemological trap,* then apparently the danger is real. As a less trivial example, let us examine how our author uses the concept of chance in his own domain, molecular biology. Unfortunately, the same methodological confusion reappears. Let us open the book on page 110. The topic is proteins, those molecules which are "formed by a linear polymerization of bodies called amino-acids." These amino-acids, numbering 20 in all, follow each other in random order along this "polypeptide chain", and the only model compatible with what we know to date about the formation of the chain is that of generalized repetitive trials. To be precise, the frequency (or, in the model, the probability) that a particular acid be found at one point of the chain depends neither on the position of the point in the chain, nor on the nature of the acids occupying neighbouring positions. In probabilistic terms, the model satisfies the condition of stationarity along the chain, as well as the condition of stochastic independence of events involving different points of the chain. But let us rather quote the author:

> "It must be emphasized that the statement that the sequence of amino-acids in a polypeptide is due to chance is not in any way equivalent to an admission of ignorance. It expresses a factual finding: namely that, for example, the mean frequency with which a particular residue (=acid) is followed by another in polypeptides is equal to the *product* of the mean frequencies of occurrence of each of the two residues in proteins in general."

This version of the formulation is totally unassailable. It does not overstep in any way the operational limits of the model, and appeals only to actually observed facts. Unfortunately, it is immediately preceded by another formulation which is very far from being equivalent to it:

> "These structures (i.e. the polypeptides) are "due to chance" in the sense that if we know exactly the order of 199 residues in a protein that contains 200, it is impossible to state any rule, whether theoretical or empirical, which would enable us to predict the nature of the one residue not yet identified by analysis."

Not only are those two formulations far from equivalent, they are actually incompatible. They refute each other, at least if we consider them as empirical formulations. In fact, the first one affirms that a certain well-defined probabilistic model, namely that of generalized repetitive trials, has not, to date, been disproved by the data. But to verify a probabilistic model, the usual method is to apply statistical tests: one chooses (before, of course looking at the data, or, at least, without being influenced by them) an event which has, in the model, a very small probability. The model is rejected if the event actually occurs. There is always some equivocation in this approach, because after all, a very improbable event may occur. But the smaller the probability of the event in question, the more is the ambiguity reduced. In particular if a previously defined event which has, in the model a probability less than 10^{-10} or even 10^{-100} were to be realized, everybody would agree that the model is refuted. For otherwise it would be simply impossible to use probabilistic models in the domain of empirical science, there being no criteria for verifying them.

As a test of the case at hand, let us choose the following event, which we shall denote for short by A: "Among distinct proteins whose chemical formula is actually known, there are at least two which have the same initial sequence of 199 amino-acids." Let us roughly evaluate the probability P(A) of this event in the model of repetitive generalized trials. I do not know the exact number of distinct proteins whose chemical formula is known at present. Suppose, to be on the safe side, that there are a hundred thousand, i.e. $n = 10^5$, and suppose that the frequencies of all 20 amino-acids are equal to 1/20. (In reality, different acids are not equiprobable, but an exact evaluation, based on observed empirical frequencies, would lead to the same orders of magnitude.) The probability that a given molecule starts with a given sequence of 199 acids is therefore, in the model $(20)^{-199}$ or approximately 10^{-258}. It is then easy to see [5] that P(A) is at most equal to $(1/2)n^2 \times 10^{-258}$, so that, certainly

$$P(A) \leqq 10^{-248} .$$

This probability is inconceivably small. If that event were to be realized, it would provide the best possible reason for rejecting a probabilistic model.

But Monod's second formulation, if we interpret it as an *empirical* observation, and not simply as the result of a theoretical deduction from the model, which has received no experimental verification – implies that this event A, of extraordinarily small probability, has in fact been realized. For we cannot maintain that it is an experimentally proven fact that there exists *no* theoretical or *empirical* rule which would enable us to predict the 200th acid, knowing the first 199, if we have not actually observed at least once two proteins starting with the same sequence of 199 acids and differing from each other only from the 200th residue onwards. But this implies that A has been realized, and we are therefore led to reject the probabilistic model.

One can go further, and show that the probabilistic model does not exclude, but on the contrary *implies* the existence of an *empirical* rule which enables us to predict the 200th acid from the first 199. Knowing that the number of distinct proteins in higher animals is of the order of one million, and assuming that in our galaxy there are at most one billion planets which have supported each one billion different species, we can take the number of proteins which exist or have existed in our galaxy to be 10^{24}. (This in undoubtedly an overvaluation.) But the probability (in the model) that, among10^{24} distinct molecules, there are two which have the same initial sequence of 199 residues is at most $(1/2)(10)^{48}/(20)^{199}$, that is about 10^{-210}. This is such a small probability that the corresponding event must be considered impossible. In other words we can affirm with a risk of error less than 10^{-210} that knowledge of the first 199 residues allows us to identify positively and unequivocally any of the proteins which exist or have ever existed in our universe. This means that there does exist a rule (albeit a purely empirical one) which allows us to predict without any risk of error not only the 200th residue, but all the others as well, from the first 199.

The nature of that rule is not veiled by any mystery. It can be compared to citizen identification. Knowledge of a small number of parameters (surname, first name, date and place of birth) allows us, with a negligible risk of error, to identify each of the 50 million citizens of our country. If, as we are threatened, the biographical data relating to each one of us were stored in the memory of some computer, anyone who had access to the super-computer would be able to predict with complete certainty the height, profession etc., of any French citizen from the knowledge of his citizen identification. Let us not hasten to object that such an operational procedure is foreign to science. For all systematic classifications in botany or zoology proceed in a way which is not fundamentally different from the above.

In the case of proteins, the rule which would enable us to perform the prediction is at present, and will undoubtedly remain forever, out of our reach. This is because it is certain that we shall never know the exact formula of the 10^{24} proteins. But this does not change the fact that the rule exists. For otherwise Monod's probabilistic model would be refuted. In any case, on the basis of the few hundreds or thousands of proteins whose structure we actually know, some form of prediction is possible: if, when we analyse a new kind of protein, we find that the first 199 residues coincide with those of an already known molecule, we can bet with complete certainty (if the model is correct) that it is the same protein, and we can then predict all the residues which have not yet been analyzed.

An Apparent Paradox

The reader may have been disturbed by the following apparent paradox: in spite of, or, more precisely, *because* of the "purely random" character of the model used to describe the succession of amino-acids along the polypeptide chain, we are able to affirm, with a degree of certainty which is rarely reached elsewhere (a probability of error at most equal to 10^{-210}!) that there exists a rule which enables us to predict the 200th residue when we know the preceding ones. We might say that we have here only a very complicated empirical rule, which we can only make explicit in the form of an exhaustive inventory, and which cannot be accounted for by any simple theory. But is this a valid objection? No one, (Monod less than anyone else), has required that this rule should be "simple." Apart from the fact that it is not easy to define precisely what we mean exactly here by simplicity, this rule, since it exists, is what it is: That it does not appear to us to be simple is a value judgement on our part, that is, a purely subjective appraisal which does not detract in any way from the objectivity of the rule in question. After all, by what right can we require the cosmos to conform to our mental structures or our aesthetic preferences?

It is actually not at all certain that the rule in question is not "simple," and we would find it very difficult to prove experimentally that it is not. For example, let us put the 20 amino-acids in some fixed order. (There are $20! = 2.4 \times 10^{18}$ ways of doing this.) To each acid there now corresponds a well-defined number between 0 and 19. The following rule [let us call it (R)]

$$a_{200} = a_1 + a_2 + \dots + a_{199} \pmod{20} \qquad \text{(R)}$$

can certainly be said to be simple. This rule might appear to us to be a priori implausible, because we could say that there is "no reason" for things to happen in this way: I agree. But in order to prove that this rule is false, we would have to try out all its 2.4×10^{18} possible versions and produce for each an experimental counter-example. No one has done this and it is certain that no one will ever do it. Suppose that a counter-example could be found every thousandth of a second on the average. We would still need a million centuries to complete the verification. Moreover, one can obviously consider many other types of "simple" rules. It could be objected that no sensible person would think in our days of studying the possibility of such "absurd" rules, i.e. rules having no intelligible connection with the present biological context. I would personally not advise a budding researcher to engage in a path which is apparently so sterile. But let us not forget that it is only for us that there is no intelligible connection between this rule and the phenomenon studied. Moreover, the common sense which makes us reject such a hypothesis today without further examination is of relatively recent date. For centuries humanity has taken delight in stranger arithmological speculations.

To enliven the paradox, let us also note that the application of rule (R) leads to numerical sequences which could be easily mistaken (except for some very few particular choices of the initial sequence e.g. $0 = a_1 = a_2 = \ldots = a_{199}$) for "purely random" sequences, i.e. sequences obtained by independent draws. More precisely, it is easy to prove the following theorem: If $A_1, A_2, \ldots, A_{199}$ are 199 independent random variables uniformly distributed on $(0,1,\ldots,19)$, then $A_{200} = A_1 + A_2 + \ldots + A_{199}$ (mod 20) is itself a random variable uniformly distributed over the first 20 integers and moreover the 199 variables $A_2, A_3, \ldots, A_{200}$ are also mutually independent. It is only at order 200, i.e. when we make explicit use of the joint distribution of the 200 variables $A_1, \ldots, A_{200}$, that it is possible to exhibit the effect of the functional dependence. In other words, even when we have a very large number (e.g. $n = 10^{24}$) of sequences of 200 numerical values obtained by the above process, the usual statistical tests will regularly and unfailingly confirm a complete absence of dependence.

It is on analogous procedures that simulation techniques, well-known to all those who have had the occasion to use in practice probability theory, are based. Contrary to what one might think, it is not at all easy to choose a number "at random" between, say, 1 and 10^6, and it is even less easy to perform a large number of such draws which would be independent of each other. This technical difficulty, which is well known to practitioners, should make us more circumspect in handling concepts such as chance which appear to us falsely to be "clear." Actually, it would be showing lack of caution to extrapolate to the cosmic scale a concept which we are already incapable of keeping in check in a truly operative way at as modest a scale as that of the simulation of a few numerical values. The way it is done in practice is completely paradoxical (and by the same token very instructive). The methods used in general by practitioners are, in fact, of a purely arithmetic nature, and *totally deterministic*. They are of the same kind, in that sense, as rule (R). For example, one can choose "any" initial numer n_1 between 0 and 10^6, raise it to the third power and retain the six middle figures of the decimal expression of the number obtained: n_2 is the number whose decimal expression is given by these six figures. The process can be repeated many thousands of times, and the values obtained submitted to the most severe statistical tests. In general, a satisfactory confirmation of the "independence" of these pseudorandom variables is obtained, except, sometimes, for certain particular values of the initial number n_1. But the process of generation of these numbers "at random" is completely deterministic. Here we are sure that there exists a simple "theoretical rule" which permits the prediction of the thousandth number from the first one, since we ourselves have chosen it. The "random" appearance of the result, which withstands the most severe tests, has its source, without any doubt, in the bizarre, heterogeneous character of the various arithmetic operations which we have performed. To raise a number to the third power and then to retain the six middle figures of its decimal expression are two arithmetic operations which have "nothing to do" with each other, at least from our viewpoint. This enables us to "understand" why numbers constructed in this way appear to be "random." But they are not, and if more powerful tests were to be performed on a very long sequence of such numbers, they would reveal their deterministic character. For example, after at most one million operations, one necessarily reverts to a numerical value already obtained, and from that point on, the sequencee becomes periodic, reproducing itself ad infinitum. We are not dealing here with an arithmetic "curiosity" but with a procedure currently used by practitioners to construct numbers "at random." The existence of a simple theoretical rule, of deterministic nature, does not in any way contradict, in their eyes, the "random," empirically corroborated, character of the numbers they obtain.

The Threshold of Realism or Objectivity

These examples illustrate well, I think, the capital difference which exists between a *theorem* and an *empirical formulation.* Any given model, however well tested and corroborated, always necessarily contains theorems which do not correspond any more to empirical formulations, which cannot be controlled, and are not even controllable, beyond a certain limit. There always exists a *threshold of realism,* beyond which a mathematician can certainly pursue happily his deductions, but which a physicist must respect, lest he obtain first uncontrolled formulations, and later uncontrollable ones, that is, formulations which lack any objective meaning. In other words, they are "metaphysical" in the sense given to this word in the usage of objective science. In the case of our proteins, for every integer n, the probability in the model of a given sequence $a_1,\ldots,a_n$ of n amino-acids is equal to the product $P(a_1)\ldots P(a_n)$ of the individual probabilities, that is, to $(20)^{-n}$ when the different acids are equiprobable. This is a theorem of the model. But if there is to correspond to it an "objective" or "empirical," i.e. experimentally controllable, formulation, then it is necessary that among the 20^n possible sequences, at least a few should be repeated enough times among the 10^{28} sequences [6] which could theoretically be observed. If we wish, for example, that there should be in the model a probability greater than 1/100 that at least one of the $(20)^{-n}$ sequences be repeated not less than ten times (and this is a minimum requirement), it is easy to see that n cannot be greater than about 24. Any formulation of a higher order, for example of order $n = 200$, like Monod's, is empirically empty [7]. If we now consider not the set of all molecules in the universe, but only those whose formula we actually know, say 10,000, this yields at most 10^8 observed configurations. We must expect that a given sequence of length n will appear, on the average, $10^8 \times 20^{-n}$ times in our experimental material. For $n = 2$ and 3, we shall have more than 10^5 and 10^4 repetitions respectively, and therefore a very good opportunity of applying experimental checks. Already for $n = 6$ the mean number is of the order of unity: at order 6 it will only be possible to check very few sequences, those which, "by chance," will have repeated themselves say at least ten times. Above $n = 7$ or 8, the experimental material available today does not allow us to effect any practical checks (or if it does allow them, this by itself means that the probabilistic model is refuted).

In conclusion. Any assertion relating to independence of order $n = 10$ cannot be considered (if the model fits) as experimentally established today. As for the concept of independence of order $n \geqq 30$ (and a fortiori of order 200) it simply lacks any objective meaning. How then can we extrapolate to the cosmic scale, and assert categorically, as if this were an experimentally proven fact, or even simply a formulation possessing some objective meaning, that "chance is the only possible source of all creation and all novelty in the biosphere?" It is likely that J. Monod conceived his philosophy as being before all a war machine against that of Teilhard de Chardin [8]. This explains their kinship. Monod's "chance" is the brother foe of the Omega point of the good father: its foe, certainly, but fundamentally its brother: they come from the same family.

Notes

[1] p. 127.
[2] Underlined by me.
[3] Underlined by J. Monod.
[4] See e.g. "Les concepts physiques" in "Logique et connaissance scientifique" by J. Piaget et al. Paris, 1967, La Pleiade.
[5] Translator's note. There are $n(n-1)/2$ pairs, and the probability that both members of a pair have the same initial sequence is 10^{-258}. The probability that at least one pair has this property is not greater than the sum of the probabilities for all pairs, namely $10^{-258} \times n(n-1)/2$, which is less than $(1/2)n^2 \times 10^{-258}$.

[6] Supposing that proteins contain on the average 10^4 acids, the 10^{24} molecules in the universe would contain approximately $10^4 \times 10^{24} = 10^{28}$ sequences of 200 consecutive acids.

[7] Let us draw in passing the attention of the logicians to the strange status (which reminds us, in a more complex form, of the famous Paradox of the Liar) of the formulations obtained, in the fashion of Monod's, by grossly overstepping the *threshold of realism* of a probabilistic model. They are *both theorems and virtual falsifiers of the model.* If they are empirically true, they are theoretically false. (i.e.: if they are experimentally verified, or even simply if the conditions necessary for an experimental check are realized, the model from which they are deduced is ipso facto refuted.) If they are theoretically true, then they are empirically unverifiable. (If the model can be thought satisfactory, i.e. not refuted by experiment, then the conditions necessary for an experimental check are not realized.) This is why they can be said to be *objectively empty*. More precisely, they can only have objective meaning if the model from which they were deduced is refuted by experiments.

[8] Monod's feelings about Teilhard de Chardin can, in my view, be described by the phrase: "disappointed love, turned into hate." Althusser has noted the frequent Teilhardian reminiscences which dot Monod's writings. He has analysed (unfortunately from the rather rigid language and viewpoint of a particular School) the ambiguous relationship which exists between the two authors. A good analysis (in normal language, and from a less particularistic viewpoint) as well as a good bibliography, will be found in Madeleine Barthelemy-Madaule's work "The ideology of Chance and Necessity" (Paris 1977). The latter author, an academic philosopher, seems to me to have well grasped the essential point: namely that the illustrious biologist, when he uses words like "chance" or "probability," simply does not know what he is talking about. But as she is not herself a probabilist, she did not think she had the authority to say this so brutally. Monod's "model" has in any case been seriously challenged on its content (i.e. from the point of view of biology) by the biologists themselves. I cannot, because of my lack of competence, take sides in that debate, and can only refer the reader to Ruffie's account in "From biology to Culture" (Paris 1976, pp. 205 ff.). The reader will be aware that my only goal in the preceding pages has been to illustrate by a striking example the dangers of extrapolating probabilistic concepts or models to infinity.

Chapter 2

Why we do not Agree with the Etruscans or On the Objectivity of Probabilistic Models

"This is where we do not agree with the Etruscans, who are specialists in the interpretation of thunderbolts. According to us, it is because of a collision of clouds that a thunderbolt explodes. According to them, collision occurs in order that the explosion should take place. As they ascribe everything to the divinity, they are convinced not that thunderbolts forecast the future because they have come into being, but that they have come into being in order to forecast the future."

Seneca (Natural Questions II,32)

The Problem

In the preceding pages, we have witnessed the overstepping of what we have called the *threshold of objectivity* of a probabilistic model, beyond which deduction yields only empirically empty statements. But does there really exist such a threshold, within which a probabilistic model is capable of having an objective meaning? We know that there exists one school of thought that disputes this: the "subjectivists" [1], who advocate a purely subjective interpretation of probability, holding that it is impossible to give an objective meaning to a probability statement concerning a unique, singular event. If I say, for example, "there is one chance in two that an intelligent form of life exists on the planet Mars", there is nothing more in this statement than a personal, purely subjective opinion. For if someone contradicts me by objecting: "Surely not. The probability is 7/10 (or 3/10 etc...)", there is no way to decide who is right. It is likely that we shall know in the near future whether Mars is "inhabited" or not, but this experimental observation will neither confirm nor refute any of our probabilistic statements, which will remain for ever undecidable. This argument cannot be faulted. But does it necessarily follow that probability can never, in any context, represent anything else but a "purely" subjective opinion? The objectivists reply that they are not interested in any way in the prediction of a unique and singular event. It is, for them, a question foreign to science. To quote K. R. Popper [2]: "Every controversy relating to the question of whether or not there are events which occur only once and cannot be repeated, is a controversy which cannot be resolved by science: it is a metaphysical controversy." What interests them is only the statistical behaviour of large aggregates, and the purely objective laws which govern them. What is only a probability at the individual scale appears, thanks to the law of large numbers, in the form of a frequency that we can grasp and measure. Our probabilistic statements are therefore purely objective, since we can test them objectively, as long as we repeat the same experiment a sufficiently large number of times.

To this the Bayesians retort: "It is you who have made the judgement that it is the *same* experiment. This is a purely subjective opinion that we are not in any way obliged to share. From the position you have adopted, it follows logically that after a sufficiently large number n of experiments you will have to use the observed frequency to evaluate the (subjective) probability which you will attribute to the $(n+1)$th experiment, which has not yet been carried out. But this does not mean in any way that this probability has acquired an objective meaning." And the dialogue of the deaf can continue indefinitely.

One can perceive a tinge of extremism in the subjectivists' argument. In the face of the indisputable and resounding success of probabilistic models in such diverse domains as thermodynamics, quantum mechanics, insurance, demography etc..., in the face of the precise predictions, continuously confirmed by observation, to which they lead, it is simply unreasonable to deny their objective value. On the basis of their successes, the physicists are convinced, with good cause, of the objectivity of their "probabilities," and categorically reject the encroachments of the subjectivists. But on the other hand, the latters' arguments concerning the impossibility of giving an objective meaning to the probability of a *unique event* are apparently irrefutable. One could perhaps, like Solomon, suggest that the child be cut in halves. I mean that we should perhaps abandon unique phenomena, without recourse, to the arbitrariness of the subjectivists, and limit the domain of objectivity of probabilistic methods to "regular and repeatable" phenomena, similar to those studied by physicists. But our aim, in this work, is precisely to examine whether, and to what extent, a *unique phenomenon* can be the object of an estimation or a prediction of the basis of fragmentary information. Since most of the estimation techniques used in practise are based on probabilistic models, it is not possible to sidestep the issue.

Some Anthropomorphic Illusions

The preceding chapter has dealt adequately, I hope, with the idealistic illusion which attributes to Chance, with a capital C, some sort of cosmic responsibility. But illusions of this kind die hard, for by nature we look avidly for profound and definitive explanations, which we think may be capable of enlightening us about the mysteries of the Universe. In physics, the deterministic models used at the macroscopic scale often correspond to the intuitive representation of an action which is transmitted from one object to the next, and leads, mathematically, to one or more partial differential equations. In the case of a vibrating string, for example, it seems clear to us that the displacement of a small element situated at point x affects the neighbouring element at $x+\delta x$, pulling it along. Our intuition of movement, formed during an evolution which required millions of years, supplies us with spontaneous representations, of archetypal nature, which make us believe that we understand what really goes on. That there is there an anthropomorphic (or zoocentric) illusion is a lesson we have learned from microphysics, perhaps to our detriment, but also perhaps to our advantage, for we have thus assimilated this essential lesson: the model is never identical to reality. There are innumerable aspects of reality which will always elude it, and on the other hand the model will always contain parasitic concepts, which have no counterpart whatsoever in reality. When we deal with a unique phenomenon and a probabilistic model, that is a space (Ω, α, P) which is put in correspondence with this unique reality, the same kind of illusion incites us to say that everything happens, after all, as if the realized event had been "drawn at random" according to law P in the sample space Ω. But this is a misleadingly clear statement, and the underlying considerations supporting it are particularly inadequate. What is the mechanism of this "random choice" that we invoke, which celestial croupier, which cosmic dice player is shaking the iron dice box of necessity? This "random draw" myth, for it is one, (in the pejorative sense), is both useless and gratuitous. It is gratuitous, for even if we assume that a unique random draw had been performed, once and for all, and an element ω_0 had been selected, we would in any case have no hope of ever reconstructing either the space Ω or the probability P. For since we are dealing with a unique event, the only source of information we possess is the unique element ω_0, which was chosen at first: all the rest, the whole space Ω and the infinite ocean of possibilities it contained, will have disappeared, erased forever by this unique draw. It is useless, that is, it has no explanatory value, basically for the same reason: for the properties that we can observe in our universe are contained in this unique element ω_0, and no longer depend on anything else. Thus we can confidently ignore all the riches which slumbered in the other elements ω, those which have not been chosen. In any case, this element ω_0, ours, had doubtlessly a zero probability of being drawn, and thus our universe is "almost impossible": nevertheless it is the only universe given to us, and the only one we can study.

The Popperian Criterion of Objectivity

Before starting the discussion, it is necessary to define precisely the meaning of the terms used. We are not attempting here a deep analysis of the relationship between subject and object [3]; we are trying to give a criterion, which should be as simple as possible, but which will enable us to recognize without error whether, and to what extent, a statement, a model, etc... will have an objective meaning, or whether it must be considered as "purely subjective." Let us imagine the following dialogue [4] between two protagonists A and B:

A: It is Jupiter who throws thunderbolts.

B: Of course not. A thunderbolt occurs when two clouds electrically charged with opposite signs meet.

A: Naturally. But this is precisely the means utilized by Jupiter when he wants to throw a thunderbolt. He makes the two clouds meet.

Or (in a version of more "modern" appearance)

A: Thunderbolts are due to Chance.

B: Of course not, etc...

A: Naturally, but it is precisely Chance that makes the two clouds meet.

The argument can continue indefinitely. To every attempt at refutation by B, A opposes an absolutely unanswerable *ad hoc* objection. For it is quite impossible to prove that it is not Jupiter (or Chance etc...) that throws thunderbolts. This is because we cannot imagine an experiment or an observation etc... whose eventual result could refute A. Thus we judge A's opinion to be "purely subjective," i.e. without any objectivity. In Fact A's opinion is compatible with everything as well as with its opposite, and conversely it can predict nothing, which means that it does not supply us with any real information.

In conclusion, the objectivity of a statement, an opinion etc... depends on the possibility of testing its correctness. According to K. R. Popper [5], "the objectivity of scientific statements resides in the fact that they can be submitted to intersubjective tests." One must however distinguish between two rather different cases, depending on whether we are dealing with singular of universal statements.

Examples of the "singular" type are statements which are "factual reports," e.g. "It is raining (here and now)", or "Allied troops landed on 6 June 1944 on the beaches of Normandy." I am well aware of the fact that there do not exist "immediate data" or "raw facts," and that what we refer to by these words always results from a preliminary elaboration and construction. But in practice, (in everyday life as well as in the daily practice of scientific work) this remark is of little interest. Let us be content with saying that, by hypothesis or by convention (according to one's taste), we shall assume that a certain consensus has been reached regarding the meaning of words and their appropriateness to describe (what we call) reality. Two observers present at the same moment at the same place will agree to say that it rains (or it does not). Two historians, having access to all sources, will agree that the landing did occur at that date. Thus, as far as singular statements are concerned, i.e. as far as "findings" concerning a past, present or future fact are concerned, the criterion of objectivity resides in the fact that once all the necessary documentation has been assembled, a consensus will be reached among "sensible" people concerning the truth or falsity of the statement. We are dealing here with "decidable" statements, i.e. with statements for which it is possible to decide unequivocally whether they are true or false as long as we have at our disposal the required information. It is possible that a decidable statement (such as: "it rained in Paris on July 1st of the year 251 B.C.") will remain, in fact, undecided because of lack of access to sources. But this does not detract in any way from its objectivity. The latter, similarly, is independent of the question of whether, in the end, the statement will turn out to be true or false. An objective statement can very well turn out to be false. It is enough that it should be possible to find it false.

We now come to statements of the universal type. We meet them, of course, in the so-called positive sciences (for example "two bodies attract each other in proportion to the inverse square of their distance"), but also in daily life ("it rains every day in London"). They refer to an unlimited class, not necessarily infinite, but of indefinite extension, e.g. all bodies which exist, have existed, and will exist in the universe. They affirm that all the elements of this class posses a property which is (at least in principle) testable. They cannot be verified (i.e. found true or false) because it is not possible to carry out an exhaustive inventory of the universe. But one can falsify them (show that they are false) by exhibiting counter-examples (for example witnesses may assure us that it did not rain in London on July 1st of last year). Their objectivity is tied to their falsifiability. In science, when we deal with a statement, a model or a theory (inasmuch as they relate to a well-defined sector of the world we call real), we say that they have an objective meaning if, and to the extent that, it is possible to submit them to test by means of experiments or observations whose result can be interpreted unequivocally. (This means that it must be possible to achieve the consensus of specialists regarding the interpretation of these results.) Very often, a scientific statement will be of universal type and therefore not susceptible of rigorous (logical) verification, because this would effectively require the performance of an infinity of observations or experiments. On the other hand, it must always be possible to deduce from a general scientific statement some more particular statements (such as prediction of the result of experiments or observations carried out under well-defined conditions) which can be either confirmed or disconfirmed (verified or falsified). If they are confirmed, it does not follow that the general statement is verified, But only that it is "corroborated" (not refuted). On the other hand, if one of them is disconfirmed, the general statement is ipso facto falsified (refuted). In other words, the general statement has an objective meaning to the extent that it is falsifiable, and possesses validity (in a relative and always provisional sense) only to the extent that it has been corroborated, that is, that it has victoriously withstood all the attempts at falsification to which it has been submitted until now. And the degree of validity which we shall attribute to it will be the higher, the more numerous and severe will have been these attempts. This is the criterion of *falsifiability* proposed by K. R. Popper [5] as a *demarcation* line between "metaphysical" and "empirical" or objective statements. It is that criterion which we shall use.

Operational Concepts

The quest for objectivity has led physicists (who can be reliable guides for us in this matter), already a long time ago, to only allow the use of *operational concepts* (in the sense of Bridgman), i.e., according to J. Ullmo [6], concepts which are "defined by the regular, repeatable process which allows us to reach them and measure them." This does not mean, or at least does not only mean, that their definition must be based on criteria which will allow us to carry out measurements; but more profoundly that the concept is defined or *constituted* by the very system of "repeatable relations" which allows us to focus on it and by the physical laws which summarize the system. Thus, electrical resistance is defined by Ohm's law $V = RI$. There is therefore, in the notion of operational concept, nothing more (but also nothing less) than a system of operations, which can actually be carried out by the physicist, and which control and confirm each other. And it is precisely the invariants which are brought to light by these mutual controls and confirmations which constitute the operational concepts.

It follows from this that the value, and even the objective meaning, of an operational concept are always relative and limited: they are relative to the scale and the domain of reality within which the operations which constitute it have a meaning. They are limited by the precision of the measurements which define it. And it is not enough to say that we can never have anything other than an approximate knowledge of the "real" values, which do actually exist, even though we do not know them. Concepts which are perfectly operational at our scale, such as length or speed, become blurred and fuzzy and, little by little, lose all objective meaning as we descend towards microscopic scales.[7] It is here that the necessity to distinguish between the model and reality acquires all its importance. For once operational concepts and the physical laws which support them have been brought together within the framework of a mathematical model, there is a great temptation to forget these limits and, blindly relying on the mathematical formalism, draw from the model conclusions which are well outside the domain of objective validity. We have seen in the preceding chapter such an example of overstepping the *threshold of objectivity*. The existence of this threshold, and the temptation to cross it, constitute a permanent danger that we must particularly keep in mind when we use probabilistic models.

Subjectivity

We shall not dwell at such length on the notion of subjectivity, insofar as it refers to the opinions, beliefs, and feelings of conviction of this or that individual. Let us mainly note that it is not in any way the logical opposite of objectivity. People said to be "reasonable" or "sensible" will often give their (subjective) agreement to a well-corroborated (objective) statement such as: "when an apple becomes detached from a tree, it falls down and does not fly towards the stars." In that sense, obviously, any probabilistic statement, insofar as some individual expresses his support for it, can always be said to be subjective. But this does not exclude a priori its objectivity. An objective law, such as the law of universal attraction, insofar as I believe it to be "true" can also be said to be subjective, since it does, in fact, represent my personal opinion. Even though the subjectivists sometimes play on words, it is obviously not what they mean. When they say that "probability" is subjective, they mean "purely subjective," in the sense of lacking any objective meaning. They do not, however, imply any connotation of arbitrariness or irrational whim. They rather refer to the undeniable fact that two different individuals, placed in the same situation and possessing the same information, may well behave in a different manner, without this implying that their behaviour is irrational. It is simply their tastes or their aspirations which differ, as well as their personal assessment of the situation (which, according to the Bayesians, can always be explicitly stated in the form of an evaluation of the subjective probability attributed to the various eventualities).

There is no Probability in Itself. There are Only Probabilistic Models

Let us now tackle our problem. We shall first reject any final assertion concerning the inherent objectivity or subjectivity of "the" probability of an event. The singular form of the word "probability" and the definite article are, strictly speaking, nonsensical. For on a non-trivial [8] sigma algebra one can always construct an infinity of probabilities. Moreover, in applications, there are many ways (all reasonable) to probabilize a given phenomenon, depending on the viewpoint adopted, the aim pursued, etc... Finally, the notion of probability is of a mathematical, not empirical nature, and thus the notion of objectivity as it is accepted in the positive sciences [9] is not relevant to it. It will of course be objected that what is being debated here is not the theory of probability as such, but its "application to reality." Unfortunately, nobody has ever applied either the theory of probability, or for that matter any other mathematical theory, to reality. One can only "apply" to reality real (physical, technical, etc.) operations, not mathematical operations. The latter only apply to mathematical models of the same nature as themselves. In other words, it is always to *probabilistic models,* and only to them, that we apply the theory of probability. And the question is to examine whether these probabilistic models may, or may not, have an objective meaning. I am not contesting that it may be possible, and sometimes even useful, for an individual to put some order in his ideas or opinions by representing them in the form of a probabilistic model [10]. But does this imply that these models, or any other we might make up by other means, are necessarily devoid of objectivity?

In conclusion. *There is no probability in itself. There are only probabilistic models. The only question that really matters, in each particular case, is whether this or that probabilistic model, in relation to this or that real phenomenon, has or has not an objective meaning. As we have seen, this is equivalent to asking whether the model is falsifiable.* Practice shows that the answer to that question may be positive. There are, in fact, cases when everybody, in the light of experimental results, will agree to abandon the probabilistic model which had been proposed at the outset. But what is a probabilistic model?

Probabilistic Models

In general terms, a "probabilistic model" consists of a "probabilized space" (Ω, α, P), *together with a convention* which enables us to establish a correspondence between the elements of the space and a certain sector of reality. The elements ω of Ω represent those states of the phenomenon or phenomena we are attempting to describe which are considered *possible*. The elements A of the sigma algebra α (traditionally called "events," but only in the sense of possible, not necessarily realized, events) represent those statements which are considered as *decidable,* i.e. those (virtual) findings which may appear as the eventual result of observations or experiments which we can carry out in relation to the phenomenon at hand. Finally the probability P is a function defined on α which attributes to every event A belonging to α a number P(A), called the numerical probability of A, and which we can, at the outset, *choose* as we wish, as long as it satisfies the axioms. In practice, two cases arise: sometimes we choose a unique and well-defined probability P from the beginning, and we then say that the model is *completely specified;* sometimes, however, we retain some room for manoeuvring by just choosing a family $P(\lambda, \mu)$ of probabilities depending on a small number of parameters λ, μ... In the latter case, we say that we have only chosen the *type* of model, and that we have left the problem of its specification, that is of the choice of the numerical values which should be attributed to the parameters λ, μ... to a later stage. According to the viewpoint of "orthodox" statisticians the choice of the type of model constitutes a "hypothesis," while the problem of the specification of the numerical values of the parameters is called "statistical inference," or "estimation" of these parameters. Since this terminology carries with it implicit presuppositions [11] concerning the real and objective existence of these parameters, we shall use the neutral and purely descriptive term of *choice* (of type of model, of parameters etc...). For in fact, after all, it is always we who choose them.

The Model of Repeated Trials

Let us look at the classical example of the type of model known as *"repeated trials."* The purpose of the model is to describe a (potentially) infinite sequence of experiments [12], each of which can yield one of two results (conventionally denoted by "success" and "failure"). The traditional example is a game of heads and tails. But we could be dealing with something quite different, such as the succession, in time, of rainy days and clear days. The space Ω is made up of all the sequences $\omega=(x_1,x_2,\ldots)$, where each x_n can be either 0 or 1, according to whether the outcome of the n-th trial is (has been or will be) a success or a failure. The numerical value of x_n is evidently determined once we know the full result of the sequence of trials, that is $\omega=(x_1,x_2,\ldots)$. There exists therefore a function X_n on Ω, such that we have precisely $x_n=X_n(\omega)$: X_n is the *random variable* associated, in the model, with the result of the n-th trial. The set α of possible findings obviously includes the eventual result of each trial, that is the events $\{X_n=0\}$ and $\{X_n=1\}$, for all values of n, but also all the events which can be derived from the previous ones by means of a finite or denumerable number of logical operations. For example, the event $\{X_1=X_2=\ldots=1\}$ or $\{\omega=(1,1,\ldots)\}$ that is, an infinity of consecutive successes, is the conjunction (or product) of the events $\{X_n=1\}$ for $n=1,2,\ldots$

In this type of model (repeated trials), the variables X_n are, *by definition* [13], independent, and, also by definition, the numerical value of the probability of success is the same for all trials, i.e.

$$P(X_1=1)=\ldots=P(X_n=1)=\ldots=p\,.$$

This common value p is the unique *parameter* on which *this type of model* depends. If we choose in advance the numerical value of p, say $p=1/2$ (which we would probably do in the case of a game of heads or tails) the model is (mathematically) completely specified. But we do not have to do this, and we may choose to wait until we have collected a certain amount of experimental information before we complete the *specification* of our model. We may even not need (as we shall see in a subsequent chapter) to make that ultimate choice. The essential point, from the point of view of methodology, is to carefully distinguish between *the two completely different roles we attribute to the same symbol p.* On the one hand, p is a parameter of the model. On account of this, it may (or may not) have an objective meaning, and the assertion "$p=1/2$" may (or may not) be falsifiable, that is, objective or empirical. On the other hand, the same symbol appears in the equality $P(X_{10}=1)=p$. Here we are stating that the probability of success at the 10-th throw (of the present game) has a fixed numerical value, e.g. $p=1/2$: a singular and undecidable statement, which is therefore certainly devoid of any objective meaning, like all statements relating to the probability of a unique event [14].

A Quest for a Criterion of Objectivity

It remains now to agree [15] on a criterion of objectivity. This criterion is strongly suggested by both intuition and terminology. In fact, even in purely mathematical works, probabilists call events of zero probability "almost impossible", and events of probability one "almost sure." The word "almost" indicates that we are not dealing with a logical impossibility or certainty. For example, in the case of repetitive trials, the event $\{X_n = 1 \text{ for all } n\}$, that is, an infinite sequence of successes without any failures, is logically possible, but has zero probability, as long as p is strictly less than one. It is therefore "almost impossible" but not impossible. In fact if we were to observe an infinite sequence of successes (say, in practice, a few thousands), we would rather choose a deterministic model of the type "one always wins at that game." It would be a perfectly objective model, (since it is falsifiable), which we might abandon later on, if we were to observe some failures. But it would certainly be preferable, given the information at our disposal, to any probabilistic model, because it is simpler. The defining criterion, (in the strict sense) for the objectivity of a probabilistic model would thus be as follows: *we agree to declare a model falsified if an event of zero probability (in the model) actually occurs (in reality)*. At this crucial point of the exposition, a few remarks are in order.

a) The first remark concerns what we have become used to call (since Cournot) the *"law of chance"*, according to which events of zero (or very low) probability never occur. A. Lichnerowicz[16] expresses it as follows: "Events whose probability is very low are experimentally impossible," and comments in the following terms: "... This law ... still remains very mysterious, and it is after all only justified by the current coincidence of the theoretical consequences of the calculus of probability with the experimental facts it interprets, without it being as yet possible to penetrate completely the secret of this coincidence." It is not out of malice that I note this naive statement which a great mathematician (who is in any case not a practitioner of the calculus of probability) has let slip out, but to show how easy it is for even thee best scientific minds[17] to *be mistaken* as soon as they deal with chance and probability. In fact, there is nothing mysterious about this "law", for the simple reason that it is not at all a law, but simply a *conventional criterion* – actually the only one which enables us to falsify (reject) a probabilistic model.

b) A (theoretical) difficulty arises from the fact that there exists, in general, (in the model), a non-denumerable infinity of almost impossible events. In the case of repeated trials, it is not only an infinite sequence of successes (1,1,...), but each of the particular possible sequences $(x_1, x_2, \ldots)$ which has zero probability. Whatever the result of an infinite sequence of experiments, that result is almost impossible. Nevertheless, one of those almost impossible results must necessarily be realized[18].

This implies that one must always *choose in advance* (before knowing the data or, at any rate, without being "influenced" by them) the event (or the finite or at most denumerable number of events) of zero or very low probability which will be used as "test." There is there, undoubtedly, a certain amount of arbitrariness, which is theoretically bothersome. But it appears that in practice it is not too difficult to establish a consensus in applications.

c) Another difficulty is the following: in most cases, the almost impossible events of the model appear as *limiting cases*, and do not correspond to any really possible experimental observations (for it is not possible to carry out an infinite number of draws). In practice, therefore, statisticians choose as test an event of small, but not zero, probability ε. Should we take ε to be 10^{-2}, 10^{-6}, 10^{-9} etc...? There is a real danger of arbitrariness here. We shall return to this point later.

d) Since our criterion is conventional, the *subjectivists* remain free to reject it, which they will surely do with the iron logic which characterizes them. Here, however, their iron logic is in danger of bringing them in conflict with normal scientific practice, and with simple common sense. I hope that I shall not distort too much their way of thinking by attributing to them the following argument: first of all, they never use a completely specified model (e.g. $p=1/2$) because no one is ever completely certain that the coin has not been tampered with. They therefore use the unspecified model, but add to it an important restriction, which modifies its probabilistic nature: for them, p is not a parameter with unknown but fixed value, but a random variable to which they will attribute a (subjective) probability law, which takes into account the information of which they dispose before the beginning of the game, but which expresses after all no more than their private opinion. For example, they choose for p a uniform law on the interval (0, 1) (or any other law). However, it is only for fixed p that the variables X_n are independent. But p is not fixed, and the X_n are therefore not independent in their model. Suppose now that after a very long game, say $n=10^6$ throws, we have obtained one million consecutive successes. Given this new information, they replace the initial ("a priori") law which they attributed to p before the experiment by the conditional law taken by this variable when we have observed one million successes. In our example, this law has probability density $(n+1)p^n$, and expectation $E(p)=(n+1)/(n+2)=0.999999$ (but not 1). This means that if they were to bet on the $(n+1)$th throw, they would behave as if it had one chance in a million to be a failure. From the practical point of view, this is more or less equivalent to a certainty of success, but there remains an irreducible theoretical difference. Their model cannot be refuted, for it is, by construction, compatible with all possible experimental results. This is logical, since they refuse to attribute to the model any objective meaning. However, in such a case, as we have previously noted, the most commendable scientific attitude would be to accept (provisionally) the deterministic model: "one always wins at that game," *and to investigate why* (e.g. could the coin have two heads, assuming that heads denotes success etc...). What *disqualifies* the subjectivist interpretation here is not its lack of logic (for it is perfectly logical) but simply its lack of interest. Pushed to its limit, this logic leads us to attribute a subjective probability to each of the singular statements that we can deduce from physical laws. For example, one could evaluate the (subjective) probability that two bodies, tomorrow, at such and such a time, will not attract each other in inverse proportion to the square of their distance to be one in a thousand (or one in a million, or a billion, as we wish). It can be done, but it is not very interesting. My thesis is simply that certain probabilistic models (not necessarily all), and some of the concepts involved (not all) have an objectivity of the same kind as that which general consensus attributes to the concepts and the laws of physics.

Operational Reconstruction of Probabilistic Concepts

We have seen that it is the operational character of a concept that determines, when all is said and done, its objectivity. This means that a mathematically well-defined concept which appears in a deterministic or probabilistic model can only be declared "objective" when it has been completely redefined, or better, *reconstructed* in strictly operational terms. This is a *radical* metamorphosis, a reshaping in depth of its personality, so to say, which elevates its initial status of simple *mathematical concept* to that of *physical concept*. When one critically examines a given probabilistic model, it is most important to carry out a *sorting out* operation. This means that one must carefully distinguish between concepts which are capable of being made *operational* and those which cannot. Only the former, as well as the statements, parameters etc... which are associated with them, can be said to be objective. The other concepts, statements, parameters etc... will remain purely *conventional*. They will have a well-defined (mathematical) meaning in the model, but no uniquely observable counterpart in the real phenomenon will correspond to them. This will restrict us to use them only in a *heuristic* capacity: to suggest methods or algorithms that we would otherwise not have thought of, but not to justify our conclusions. More precisely, the conclusions that they suggest to us will have to be critically screened, reformulated in operational terms and submitted to objective tests before they can be (provisionally) adopted. The rule, here, will be *to ensure that all traces of these conventional concepts or parameters will have disappeared from the ultimate result.*

It is just in that way that the physicists proceed, in the case of repeated trials, to reconstruct the (mathematical) concept of the probability p and transform it into a very different *physical concept,* that of *theoretical frequency*. Their procedure can be described as follows: to every (finite) sequence of n trials which yields the results $(x_1,x_2,\ldots,x_n)$ we associate the mean number of successes in the sequence, i.e. the number

$$f_n=\frac{1}{n}\sum_{k=1}^{n} x_k$$

which is called the *empirical frequency*. This frequency f_n is clearly a function of the complete sequence $\omega=(x_1,x_2,\ldots)$, and is therefore a random variable of the model, which has a well-defined probability law in the model. We can then consider the infinite sequence $(f_1,f_2,\ldots)$ of the successive empirical frequencies. It may or may not converge, converge to p or to some other numerical value. The set of points ω for which the sequence f_n converges to p is an event of the model (that is, it belongs to α), namely the event $\{f_n \text{ converges to } p\}$, or $\{f_n \to p\}$. It can be proved (mathematically) that this event has, in the model, probability one:

$$P(f_n \to p)=1 \ .$$

This *theorem,* known as the *"strong law of large numbers,"* tells us that, in the model, the probability that the sequence of empirical frequencies will converge to the parameter p is equal to one.

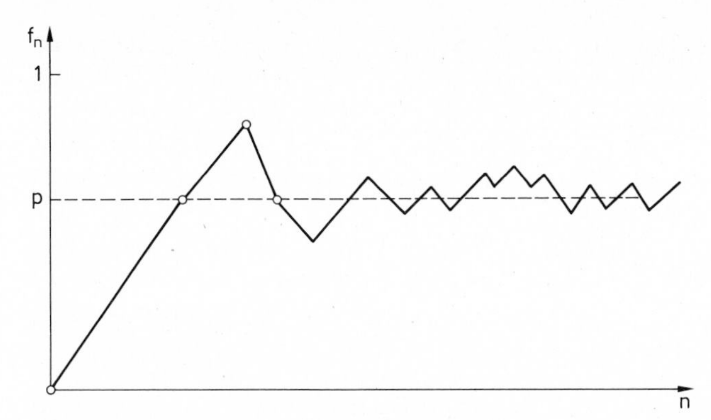

Fig. 1. Convergence of the observed frequencies

Since the event $\{f_n \to p\}$ is almost sure (in the model), we may choose it as a criterion of falsification. In other words, we shall decide in principle to reject the model if the numerical sequence of observed frequencies in the actual experiment does not converge. In practice, of course, we can only carry out a large number N of trials. This will be as large as we wish or we can, but we will have to stop eventually. For a physicist, however, the experimental results will already be enlightening. If he observes that the amplitude of the oscillations, which was originally quite large, diminishes in a more or less regular fashion, so that when n increases, the graph appears to stabilize around a horizontal asymptote of ordinate p, he will provisionally adopt the hypothesis "f_n converges to p" and will proceed to carry out various tests. He will continue the experiment beyond N, extract and study separately some subsequences [e.g. the sequence of odd terms $(f_1,f_3,\ldots)$ and the sequence of even terms $(f_2,f_4,\ldots)$], carry out further [19] experiments etc... If these various tests and mutual confirmations all highlight the existence of the same asymptotic limit p, they confer to it by the same token the status of *operational concept*. The mathematical concept is now replaced by the physical concept so constructed, to which one can give the name *theoretical frequency*. To this concept we associate the numerical value of the asymptotic limit which was brought to light by the very same process which has enabled us to construct the concept, and we also denote by p the result of this measurement. In reality, p is only known up to a certain approximation, but that is the lot of all physical concepts and measures. We may, in any case, in principle, improve this approximation, if we wish, by carrying out further experiments. Moreover, the speed with which the sequence f_n converges towards p (in the model) is the object of precise theorems, whose statement may also be reconstructed in operational terms and may thus help us to get an idea of the order of magnitude of the approximation error.

Other concepts may also benefit from this metamorphosis: for example stationarity. We cannot "verify" that $P(X=1)$ is independent of n, since this is a statement which involves probabilities of individual events. But if we have carried out say $N=10{,}000$ trials, we may divide this sequence into 100 blocks of

100 consecutive trials and examine whether each of the 100 corresponding frequencies is or is not near to p. We can compare the empirical distribution of these 100 frequencies with that predicted by the model and apply precise statistical tests (suggested by the model) etc... Again, this whole procedure of tests and mutual confirmations gives birth to a new operational concept, that of physical stationarity.

The same applies to *independence*. We shall never be able to "verify" that X_3 and X_4 are independent (because this is not an objective statement). But, just as we have constructed the operational concept of frequency of a sequence, we can construct that of frequency of two consecutive successes, or of a success followed by a failure etc... We may thus exhibit a *physical law* according to which the frequency of two consecutive events is equal to the product of the individual frequencies, a physical law which, in turn, may serve as an operational definition of independence of order 2. One may similarly give a physical meaning to independence of order 3 or 4. But, as we have seen, we cannot go very far, and independence of order 200, for example, will remain inaccessible.

In this (ideal) case, we have succeeded in reconstructing, starting from the initial mathematical model, a completely objective *physical model*. The transposition is not complete, nor can it ever be. Nothing can ever guarantee us that trials 13 and 14 were independent and had the same probability as the others. In a game of heads and tails, it could happen that, just for these two trials, but not for any other, a two-headed coin was substituted for the original coin. Statistical analysis will never enable us to find this out. But the essential part, the objective content of our model, has been safeguarded.

The Anticipatory Hypothesis and the Risk of Radical Error

We have given ourselves an unfair advantage, in this example, by assuming that we can, in principle, carry out as many trials as we wish. This is how the physicists, our teachers in this field, proceed (or at least reason). But in practice things frequently turn out very differently. We must often make an immediate decision on the basis of limited information, which we have neither the time nor the means to complete. For example, suppose that in the previous example we have carried out 20 trials giving the following results (denoted in the sequel by S):

$$(S): 11001001000011111101 .$$

We are required, on the basis of these 20 data points alone, to make a prediction concerning the mean frequency of appearance of the number 1 in the next 10,000 trials, and if possible, to state the order of magnitude of the possible error.

An "orthodox" statistician will advance the "hypothesis" [20] that "we are dealing" with repeated trials with an unknown parameter p (in this case he will "estimate" p by the relative frequency $f = 11/20 = 0.55$) or else (if there are indications pointing to it) by $p=1/2$ for example. He will then carry out "tests," which will in this case yield positive results, or, more correctly, will not yield negative results, concerning independence of (order 2), the value $p=1/2$, etc... If he has chosen the fully specified model $p=1/2$, he will then predict, for the next 10,000 trials, a number of successes equal to $5{,}000 \pm 100$ with a "probability of error" of 5% (corresponding to 2 standard deviations). I may add that a subjectivist would give an approximately equivalent prediction, although expressed in a different terminology.

Obviously, this prediction may turn out to be radically wrong. It may happen, for example, that the nature of the phenomenon will change, for a reason we cannot anticipate, and that there will appear only zeroes after the thousandth trial. The statistician has therefore taken a *risk*. This is inevitable. But, if there is a risk, that is, if his prediction can be found false later on, this means that the prediction is falsifiable, and has therefore an objective meaning. The statistician has in fact put forward an objective (falsifiable) hypothesis: not exactly the one he has stated, which only concerns the mathematical model, but, implicitly, an *anticipatory hypothesis concerning the validity of the physical model,* whose operational construction we have explained above. It is an objective hypothesis, (since it may turn out to be false after the fact), and it is anticipatory (since the most rigorous tests carried out on the 20 data points, assuming they corroborate the validity of the model for these 20 points, will not give us any guarantee that the nature of the phenomenon will not change later). Its very objectivity (since it introduces an additional piece of information which was not contained in the original 20 data points) enables us to extract from the data more than they contain, and to put forward a prediction. But its anticipatory nature (since it was adopted before its validity was tested) introduces a risk of *radical error* [21]: and this risk is the necessary counterpart of the gain in information it provided. We shall return to that point and discuss it at length in what follows.

There are other unexpected possibilities. Suppose that we had the (bizarre) idea of putting a decimal point after the first two ones, and interpret the sequence S as the binary expression of a number $x<4$: when we transcribe x into decimal notation, we find, since there are 20 binary symbols, that x lies between 3.141590 and 3.141594. Should we conclude that, for some unknown reason, we have $x=\pi$, and put forward, as a "deterministic" prediction, that the next 10,000 trials will correspond to the symbols representing the number π in binary notation? This new anticipatory hypothesis is much stronger (and therefore much more easy to falsify) than the previous one, but is of the same nature, and, curiously, not wholly incompatible with it [22].

Panscopic and Monoscopic Models

The idea of an anticipatory hypothesis which we have been led to put forward, together with the risk of radical error which it necessarily entails, brings us to the heart of the debate which opposes "objectivists" and "subjectivists." Behind the verbal quarrels, the sectarian passions [23] and the more or less inadequate terminologies, one perceives a more fundamental conflict. It is, basically, the conflict between the ideal of "scientific knowledge" and the necessities of "real life" and "practice." The scientist attempts to construct models which are as rich, comprehensive and well corroborated as possible. He assumes that one has (at least in principle) both the opportunity and the *time* to collect all the necessary documentation, and to carry out all the desirable tests, before making a judgement regarding the objective validity of each of the hypotheses. The practitioner, on the other hand, faces situations the essence of which is *urgency*. In the heat of action, only fragmentary information, sometimes of doubtful quality, is usually available. Nevertheless, one has no choice but make important decisions on the basis of that information. This is so in the social, economic, and political domain, as well as in the professional or private life of each of us. At the cost of some exaggeration, one may contrast the objective and disinterested character of scientific knowledge with the subjectivity and savage greed which reign in the domain of practical action. But one may also exalt the life of the man of action, compared to that of the researcher, who may be accused of avoiding real life and its responsibilities. One may also, perhaps more reasonably, think that there is here a difference of degree rather than of nature. For the information available to the scientist is never perfect, and he must, therefore, put forward, under his own responsibility, hypotheses whose validity, however well corroborated they are today, may be challenged at any time by the unexpected result of a new experiment. On the other hand, the practitioner is never so completely unarmed that the anticipatory hypotheses (whether explicit or implicit) on which he bases his decisions should be totally devoid of any objective value. Otherwise the merciless process of natural selection would make him quickly disappear from the scene of practical action.

Nevertheless, there does exist a real difference between the *objectives* sought and the *criteria* used in these two types of activity. First, as far as objectives are concerned, one may notice that our sciences and our techniques seem to hesitate, or to oscillate, between a speculative or explanatory mode and an instrumental or manipulative one. The former mode, an idealistic and "disinterested" one [24], claims to aim at knowledge for itself, and to understand or explain the object as it exists, independently of us and of the practical applications we may have in mind. Not that these applications are despicable. But they should be gained, so to speak, as a "bonus," and one is only interested in them insofar as they illustrate or corroborate the explanatory model. In the second mode, one seeks to change the world rather than explain it, to manipulate and tame the object, without being interested in what it is in itself. One espouses the philosophy of "nominalism:" the model is not reality. And one also resolutely adopts "instrumentalism:" the best model is the most efficient one.

The two models are not actually irreconcilable. Manipulation is only efficient insofar as the model used conforms, in one way or another, to reality, or at least to the aspect of reality we are interested in; and the most disinterested theories are only scientific insofar as they remain operational. After all, it is only by the efficiency of the predictions they allow us to make that we may judge the degree of their validity. However, if we take into account the number, size and generality of the objectives pursued, we may say, in general, that scientific models or theories are *panscopic* [25], while it is sufficient for practical or technical models to be, in a more modest way, polyscopic, or even strictly *monoscopic*.

In fact, for a scientific model or theory to be valid, it must be capable, by definition, or predicting correctly the outcome of any experiment or observation that our present technical means and knowledge enable us to imagine and effectively carry out (at least within the domain where the theory is operational). And by the same token it must therefore enable us to solve all the practical problems which are meaningful at the present time and which we can set ourselves in this domain. Thus the criterion of validity is operational efficiency with respect to *all* the objectives which one may think of. We shall therefore say that scientific theories have a *panscopic* character.

But in order to achieve the panscopic scientific ideal, there is a real need for time, thought, means and an enormous amount of information. In practical activities, this ideal is often a luxury in which, in the heat of action, one has neither the time nor the means to indulge. This is the reason for the more or less monoscopic character of practical models, which results, so to say, from a parsimony principle. For in technical, economic and other similar activities we are constantly confronted with practical problems which it is imperative to solve. (The solution should be as good as possible, but we *must* produce one.) It is imperative to take decisions on the basis of insufficient information. Under such circumstances, we rarely have the time (or even the inclination) to query at length the objective characteristics of the situation which do not directly concern the problem to be solved or the decision to be taken. What we therefore expect, before anything else, from a model (whether probabilistic or not) is actual operational efficiency just in the domain which concerns us, and we do not care if the model is grossly wrong in relation to other aspects of the situation which are of no direct interest to us. Our objective is *monoscopic,* essentially for reasons of urgency and choice of priorities.

Validity *criteria* are evidently not the same for the two types of models. A scientific theory, which is panscopic, will be automatically refuted if any one of the consequences we can deduce from it (within the domain where it is operational) is refuted by experiment. The monoscopic model, on the other hand, must be judged on its adequacy to fulfil the unique goal it pursues. But by the same token, one cannot expect from it sensible answers to questions it was not made to handle.

Moreover, contrary to scientific theories, which always have a higher or lower degree of universality, the monoscopic model often concerns a particular phenomenon, a unique situation. We shall never again encounter an identical one. The condition of repeatability, without which it is difficult to set up operational definitions and to justify the objective validity of a model, is lacking here. Let us add to this the fact that the monoscopic model is chosen in the heat of action, that is, on the basis of information which is by definition insufficient, and that it is therefore always an *anticipatory hypothesis,* which may eventually be testable *"after the fact,"* but *whose validity is not guaranteed, strictly speaking, by anything at the time of its adoption.* It is then conceivable that one may wish to query its epistemological status, and even, like the Bayesians, its objective meaning.

We shall return at length to the important question of the meaning and objective value of *monoscopic* probabilistic models. The important point to remember at the moment is that when the monoscopic model is chosen it introduces an anticipatory hypothesis whose legitimacy can in no way be guaranteed by its compatibility with the available numerical data. This hypothesis is in fact equivalent to assuming that the structural characteristics which we have extracted from the data may be extrapolated without change to the unknown parts of the phenomenon. Or, in other words, that the phenomenon behaves, where it is not known, in a manner sufficiently analogous to that which we have observed where it is known. Two conclusions follow from this: in order to choose a hypothesis of this type, one must

(1) take into account *all* sources of information, whether numerical or not, which we have at our disposal (e.g. general knowledge about the physics of the phenomenon, experience acquired with similar cases etc...);

(2) weaken the hypothesis as much as possible, so as to reduce it to the absolute minimum which is indispensable for fulfilling the purpose of the monoscopic model.

Thus we strive for *maximal exploitation of all sources of information,* and apply a *principle of strict economy in the choice of anticipatory hypotheses.*

External Criteria and the Objectivity of a Methodology

In dealing with a specific (usually monoscopic) model, whose purpose is to represent a *unique phenomenon,* it is sometimes difficult to give a rigorous justification for the meaning and objective value of the specific model. We shall see later that there exist *internal criteria* of objectivity, which may provide at least a partial answer, even in the case of a unique phenomenon. But although it is not evident that one may be able to pass judgement in each individual case, this is not the case as far as the general *methodology* which we use to choose monoscopic models in each particular case is concerned. It will be clear, in the long run, whether the latter is, or is not, efficient. It is true that each situation, each problem is unique and is dealt with by means of an ad hoc monoscopic model. But there are classes of situtions and problems which are not identical but are sufficiently analogous to enable us to formalize, at least partially, the rules which dictate the choice of the model to be adopted. These rules will eventually form a methodological system, which will be put to the test of practical action, and will either be validated or abandoned.

In other words, it is true that, when all is said and done, it is the possibility of repetition which is the foundation of objectivity (cf. Ullmo's "repeatable relation" [6]). But this does not mean that there is no possible science of the unique. First of all, it is always in a relative sense that we speak of repeating the "same" experiment. Strictly speaking, there do not exist two identical experiments: they always differ in some accessory factors (but it is we who judge them to be accessory), and by conditions of time and space. All we can say is that the factors which appear to us to be important have been made as similar as our technical means allow us, and that the others will be what they will be. Thus, one should not speak of a reproducible phenomenon, but of a class of phenomena which we judge sufficiently near to each other to consider them as equivalent. This proximity or resemblance may actually be only qualitative or structural, without encompassing the equality of the numerical parameters which describe the object. For example in geology or astronomy there are no two identical objects. But this does not stop us from basing their objectivity on the repetition of the similar.

In the same way, the so-called "unique" objects or situations are certainly unique as individuals, but they may be classified into species or types which group qualitatively and structurally analogous objects or situations. The repeatability which is a condition of objectivity will operate within these classes and it is up to us to define them as precisely as possible, for this definition will play a constitutive role in our discipline. It therefore appears that in the controversy between frequentists and subjectivists on the interpretation of "probability," it is the frequentists who are right, but only in a wide sense. The *long-run* foundation of objectivity is the systematic comparison of probabilistic predictions with realization, over a class of experiments perhaps not completely identical, but belonging to the same type, defined in a more or less wide manner. If that objectivity is not that of a singular prediction, it will be at least that of the methodology which has led to it. If this prediction reflected nothing more than the frame of mind of the practitioner or the decision maker, it would be of no interest to anyone, not even the decision maker himself. What he really wants is that the decision should be as efficient as possible, and this efficiency clearly depends on the objective situation as well.

In conclusion, the *external criterion of objectivity* is whether or not it is *"right on the average"* to use such and such a *methodology* to attempt to solve such and such a category of problems.

In order to formulate this in a slightly more technical way, let us note that in order to use the model of repeated trials or analogous models [26] based on sequences of independent, identically distributed random variables, it is not necessary to assume at any stage that we are dealing (in the real world) with the same experiment which is repeated an infinite number of times. I would not go as far as to assert categorically, like the subjectivists, that the latter formulation cannot, under any circumstances, have any objective meaning. For it seems to me that physicists are, after all, people who know what they say (and what they do). But in our field of interest there would be some very real difficulties. Fortunately, it so happens that this hypothesis is in no way necessary for our criterion of external (methodological) objectivity to be applicable. In principle, we may include any experiments, as long as we can, or know how to, analyse them. Evidently using common sense is not forbidden even in scientific matters, and one would not lump together in the same model the French elections of 1978, petroleum deposits in the Middle East, the age of the captain and the goat of Mr. Seguin. But we could well put together experiments which are really very different.

It is best here to give an example. There is an extraordinary variety of types of copper deposits, some small and some large, some very rich and some with a very low average grade. Their structure may be simple or complicated, their grades weakly or strongly dispersed, etc... Moreover, the information at our disposal for estimating them can also be extremely variable as far as nature, quality and quantity is concerned. Ignoring technical details, let us assume that we can uniquely define a notion of resource or quantity Q, expressed in thousands of tons of copper, for each of the deposits considered. This quantity will only be known after the deposit has been actually mined. Suppose (and this is actually the case) that we have a methodology which enables us to integrate, within a probabilistic model, all the information which is available at the time when the deposit must be estimated (that is, just before mining starts). In the model there corresponds to Q a random variable, which we shall also denote by Q, and the model contains the statement "there is one chance in two that Q will be larger than q." Here q is a numerical value which we can calculate (in the model, q is the median of the distribution of the random variable Q, conditional on the available information). It is of no importance at the moment whether each individual statement has or has not an objective meaning. Once we have thus estimated one hundred deposits, let us number them from $n=1$ to $n=100$. After they have been mined, we can compare, for each one, the true value Q_n and the median q_n which our methodology had attributed to the corresponding variable in the model. Set $X_n=1$ if $Q_n \geqq q_n$ and $X_n=0$ if $Q_n < q_n$. We now have a sequence of successes and failures to which we associate the (fully specified) model of repeated trials with $p=1/2$. We may then test that model (more precisely test the validity of the physical model which the probabilistic model enables us to reconstruct). If we decide that 100 trials are not enough even from the practical point of view (although in the eyes of the practitioners in the mining industry they would certainly be enough), then we can wait until we have 200 or 1,000. But we can (or shall eventually be able to) pass judgement on our methodology. The latter is therefore objective since it can, and will be, rejected if it leads, on the average, to unacceptable prediction errors[27].

Criteria of Internal Objectivity Linked to the Concrete Character of the Space

According to Ullmo's article quoted above [6], the "idea of chance," meaning, in fact, the use of probabilistic models, appears in physics "when experimentally *inseparable* initial conditions are followed by a manifest *separation* of the observed phenomena." The best we can do here is to illustrate this profound concept with a simple example. Let us consider the (idealized) case of a *circular billiard table*. Suppose that the ball, which is initially knocked without any spin, bounces back from the circular cushion according to the laws of reflection, so that at the next impact the angle θ which determines its position increases by an amount $\Delta\theta$ which, in principle, remains constant.

Let us also assume that energy losses are negligible, so that the ball circulates indefinitely around the billiard table. Let us introduce a simple physical notion, the *visit frequency* $f_n(l)$ of some arc of length l measured in radians on the cushion during n consecutive bouncings. In other words, $f_n(l)=k/n$, where k is the number of occasions when the bouncing occurs at a point of the arc l. One can then prove the following theorem: as n tends to infinity, *the visit frequency tends to the ratio* $l/2\pi$, with the sole proviso that the number $\Delta\theta/2\pi$ *is not a rational number*. If, however, the number $\Delta\theta/2\pi$ is a rational number, then the ball will return an infinite number of times to a finite number of points, and the others will never be visited. This theorem suggests to us irresistibly that the *uniform probability law* must intervene, in some way or other, in the description of the behaviour of the ball, with the possible (important) exception of the case when $\Delta\theta/2\pi$ is "by chance" a rational number. We have the intuitive feeling that the number of exceptional cases must be very small, and that in practice they will never occur. If we choose "at random," that is, with a uniform density an arc $\Delta\theta$ between 0 and 2π there is, in fact, zero probability that $\Delta\theta/2\pi$ will be a rational number. For a physicist, this eventuality is excluded, since we are dealing with a macroscopic phenomenon.

Starting from position θ_0 (assumed, for simplicity, to be exactly known), the ball will occupy position $\theta_n=\theta_0+n\Delta\theta$ after n bouncings. No matter how precise

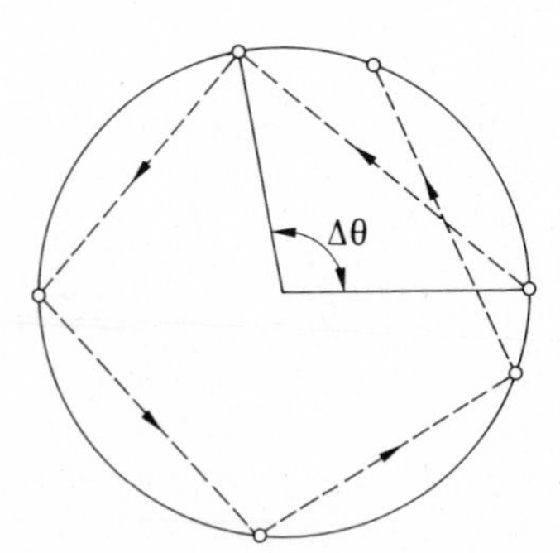

Fig. 2. The circular billiard table

our initial measurement of $\Delta\theta$, which is known, say, with an error of the order of a very small number ε, this error will become immeasurably magnified, since we can only predict θ_n up to $\pm n\varepsilon$. As soon as n is large enough, this error will be greater than 2π, or any given multiple of it, e.g. 10 times 2π. Which means that the n-th bouncing will occur anywhere on the cushion. Ullmo's *inseparable* initial conditions have been *separated* in the final outcome. This suggests a probabilistic model (which can be tested by repeating the experiment) in which θ_n is represented by a random variable uniformly distributed on the unit circle.

Independently of any experimental tests, the choice of this uniform law is inescapably imposed on us by the very physical nature of the model. In fact, supppose that we choose any law other than the uniform for θ_n (as long as it has a density [28]). One can then prove without too much trouble that for large enough m the law of θ_{n+m} will differ from the uniform law by as little as we like. It is therefore enough to replace n by n+m to obtain again the uniform law. This phenomenon of convergence to a uniquely determined limit law is a manifestation of the property known as *ergodicity*.

The simplest way to construct the probabilistic model is to take ε to be a random variable which has an arbitrary probability law, subject only to the condition that it has a density. Then $\Delta\theta$ is of the form $\Delta\theta = \Delta\theta_0 + \varepsilon$, with the value of $\Delta\theta_0$ known and, in the model, $\theta_n = \theta_0 + n\Delta\theta_0 + n\varepsilon$ is a random variable whose probability law converges as before to the uniform law. The value of this formulation is that now the exceptional case ($\Delta\theta/2\pi$ rational) has, in the model, zero probability (whatever the probability law of ε as long as it has a density).

One may further improve the model by noting that in the layout of the billiard table and its cushion there are perforce small irregularities. On account of this the successive increments $\Delta\theta_n = \theta_{n+1} - \theta_n$ cannot be exactly equal. And even if one of them happened by chance to be a rational multiple of 2π, they would not all be. The exceptional case is thus excluded by physical considerations. Let now $\Delta\theta_n = \Delta\theta + \varepsilon_n$, with the random variables ε_n (in the model) arbitrary (and not necessarily independent: it is enough that their correlation should not be too high). It can then be proved that θ_n, θ_{2n}, θ_{3n} etc... are, in the model, for sufficiently large n, *independent random variables* with the same *uniform law*. And the model can now be tested experimentally, without having to repeat the whole experiment.

I shall not dwell any longer on this example [29], whose sole aim was to show how concrete physical considerations may sometimes impose the choice of a unique and well-defined probabilistic model. It is the physical analysis of the phenomenon itself which has supplied us with criteria of *internal objectivity* which enable us to test "from the inside" the objectivity of the probabilistic model, without having recourse to repetition of the "same" experiment.

However, in the type of examples supplied by physics, we do not in fact have unique phenomena, since the condition of repeatability is always fulfilled. This allows us to reinforce and control by means of external tests based on repetition the purely internal criteria of objectivity of these models. So let us give one more example, which looks quite innocuous, but which will turn out to be highly enlightening when we shall search, in Part II, for strictly objective models, or, as we shall call them, *probabilistic representations* which will allow us to describe unique phenomena.

Let S be a bounded region of usual space (e.g. a two-dimensional space) and f(x) a function [30] defined on S. Let there also be given on S a probability law, say the uniform law on S. In the model where x denotes the random point obtained by "drawing at random" a point x in S according to the above probability law [31], the function f(x) becomes a random variable. The probability law of this variable is defined by its distribution function F. The probability that $f(x) \leqq a$, F(a), is the measure (the *area* if the space is two dimensional) of the subset of S defined by the condition $f(x) \leqq a$. This law has therefore a fully concrete meaning. It has, in particular, an objective meaning: once we know the function f, any probabilistic statement which relates, in the model, to the random variable $f(x)$ is now strictly decidable. All the parameters associated with this law also have a strictly objective meaning. For example, the expectation m of the random variable $f(x)$ of the model is, by construction, equal to the mean value in S of the function f, namely

$$m = (1/S) \int_S f(x)\,dx\,,$$

a quantity which has a perfectly objective (decidable) meaning.

Suppose now that we "choose at random" in S, independently of each other, points $x_1, x_2, \ldots$ (This can be technically done by using tables of random numbers, or by means of the "arithmetic" algorithms we have considered in the previous chapter). The numerical values we obtain, $f(x_1)$, $f(x_2)$, ... can then be interpreted within the framework of a model based on a sequence of independent random variables having all the same law F. In particular, if the function f, although perfectly determined, is not known to us, so that our only data are a finite sequence of numerical values $f(x_1)$, $f(x_2)$, ..., $f(x_n)$, we may reconstruct the conditions of the theoretical problem of "statistical inference" and of "estimation" for m within a purely concrete framework. But now we have the enormous advantage of being assured, at the outset, of the objective meaning of the law we are attempting to infer, and of the parameter m we are attempting to estimate. Moreover, some very down-to-earth considerations relating to the numerical approximation of the space integral which defines m based on a small number of points $x_1, x_2, \ldots, x_n$, will throw a crude, but appreciable light on the somewhat mysterious nature of this famous "statistical inference". Within this framework, it is permissible to talk about the estimation [32] of the parameter m without inverted commas, since m really exists outside our probabilistic model.

Notes

[1] They are also called "Bayesians", because of their intensive use of Bayes' theorem (which is of course purely classical). This is an improper appellation, for Bayes' theorem, like all other theorems, is of a purely formal (mathematical) nature, and does not prejudge in any way whether a subjective or objective interpretation will be given, in applications, to the probabilities involved. The physicists, who are objectivists by vocation, also use this theorem (e.g. in the analysis of "retrodiction", a word formed analogously to "prediction" to designate the reconstruction of the past from the present).

[2] "The Logic of Scientific Discovery", p. 43 of the French edition.

[3] On objectivity, conceived as the outcome of an elaboration or a reconstruction of reality by the (epistemic) subject, see the work of J. Piaget, and in particular "Logique et Connaissance Scientifique" (op.cit.).

[4] cf. Seneca's quotation in the epigraph of the chapter.

[5] "The Logic of Scientific Discovery".

[6] "Les concepts physiques" in "Logique et Connaissance Scientifique", op.cit.

[7] The length of a ruler may be defined, say, to the nearest millimeter, but not to fifteen correct decimal places. A fortiori, the question of whether that length can be expressed, in cm, by a rational or an irrational number has absolutely no meaning for the physicist.

[8] i.e., containing at least one event distinct from either the impossible (empty) event or the certain event (the whole space).

[9] Nobody, it seems, speaks of a subjective (or objective) vector space.

[10] It seems to me that the lottery technique used by the subjectivists to evaluate probabilities is useful in practice only when the number of random variables which are not independent is not too large. Their procedure would be difficult to implement in the case of random functions, which theoretically involve an infinity of variables (and in practice many thousands).

[11] On this point, I agree for the most part with the severe criticism levelled at that terminology by the "subjectivists". See for example B. de Finetti, Theory of Probability, J. Wiley 1974, passim. (But I do not draw the same conclusions.)

[12] I am definitely not saying that it is the "same experiment", which would have, strictly speaking, (and the subjectivists are right on that point) no objective meaning.

[13] We are talking here about the definition of a *mathematical* model, not about an assertion relating to the physical world. The subjectivists (see e.g. B. de Finetti, op.cit.) are right to underline that any assertion relating to the independence of the outcomes of the first and the second throw (in this particular game, played on such and such a day, with such and such a coin, played by Messrs. Durand and Dubois) is devoid of any objective meaning.

[14] Here again I agree with the "subjectivists".

[15] I am using this expression on purpose, because every criterion of objectivity depends, in the final analysis, on a consensus.

[16] "Remarques sur les Mathématiques et la réalité, in "Logique et Connaissance Scientifique," (op.cit.) p. 82.

[17] c.f. the example of J. Monod in the previous chapter.

[18] The union of all these events of zero probability has therefore probability one, which does not contradict the axioms, since they form a non-denumerable collection. We see that the restriction of the additivity axiom to *denumerable* collections is an essential limitation.

[19] This assumes, in principle, an extension of the original model, which should now contain variables subscripted by two indices $X_{n,m}$ (representing the n-th result of the m-th experiment), but this does not raise any difficulties.

[20] In our terminology, this is not a hypothesis, but a choice of type of model, with possibly the full specification $p=1/2$. But this choice does imply, as we shall see, an objective (falsifiable) hypothesis.

[21] The order of magnitude of its size is different from that of the "5% confidence interval" of ± 100 which is predicted by the model. This why I call it a *radical* error.

[22] In the sense that the usual tests, when applied to the sequence of the first n symbols of the binary expression of the number π, would probably not reject the model of "repetitive trials." The event itself "$(x_1, x_2, \ldots, x_n)$ = the first n binary symbols of π" has in the model, the probability $1/2^n$, already quite small for $n=20$ (about one millionth), but cannot legitimately be used as a criterion, since its choice was suggested by the data. In fact, any sequence of 20 symbols can be considered as the beginning of the binary representation of some more or less remarkable number (such as integral multiples or powers of π etc...: all we need to do is find a million of them).

[23] One finds, in the writings of some subjectivists, sensational proclamations, concerning for example the 21st century, which will apparently be a "Bayesian" century. Even an author who usually expresses himself in a very sensible and courteous manner, like B. de Finetti, sometimes lets a final ukase slip out. For example, we read in his "Theory of Probability" (J. Wiley and Sons 1974) p. 190: "Speaking of unknown probabilities must be forbidden as meaningless." Why this stamping of hobnailed boots in the republic of scientists?

[24] "Disinterested action is, in reality, *very* interesting, and interested, even admitting that ..." notes Nietzsche (Beyond good and evil).

[25] Panscopic: all objectives; monoscopic: only one objective.

[26] The most general class of this type of models is given by a probability space (Ω, α, P) defined as the product of an infinite sequence of spaces $(\Omega_n, \alpha_n, P_n)$ identical to each other (in rigorous terms: isomorphic). Each factor space describes one given phenomenon or experiment. The product space corresponds to the whole set of phenomena or experiments. Since the probability P is defined as the product of the probabilities P_n, the variables X_n, Y_n etc. describing the n-th experiment are, by construction, independent (in the model) of the variables which appear in all the other experiments. This class of models is the mathematical counterpart of the physical notion of *repeatability*. Note the *formal, or conventional, character of the space* Ω (simply defined, after all, as the set of all the possible results of all the experiments we can imagine) and its *modular* character (in the sense that it can be *extended at will*): one can always enrich it at will by adding further isomorphic factors $(\Omega_n, \alpha_n, P_n)$ each time there is a need to take into account a new sequence of possible experiments. This is in contrast with the *concrete character of the space* Ω in models where there exist (as is the case in physics) *internal criteria of objectivity*.

[27] One can go further. To each Q_n our methodology associates the probability law F_n of the random variable which corresponds to it in the model. It is well known that, in the model, $F_n(Q_n)$ is then a random variable uniformly distributed on (0, 1). One can therefore use the sequence $X_n = F_n(Q_n)$ of numerical values obtained to test the validity of the model: "X_n $(n=1,2,\ldots)$ is a sequence of independent random variables, each uniformly distributed on (0, 1)".

[28] From the point of view of physics, this restriction is not bothersome. For in order to attribute a concentration of probability to a point of known position, it would be necessary to have extraordinarily precise information on the error ε and this is experimentally impossible.

[29] Statistical Mechanics would supply, in a more grandiose setting, similar examples, and would lead us to conclusions of the same nature: it is the concrete character of the model or of the physical theory which unequivocally imposes on us here the choice of the probabilistic model.

[30] The function must obviously be measurable.

[31] In more precise language, we are dealing here with a model (Ω, α, P) where $\Omega = S$, α is the family of Borel subsets of S and P is the uniform probability on S. In concrete

terms, S could represent a mineral deposit, f(x) the grade at point x, and x a sample (for example a drilling) made at *random* in S.

[32] I remind the reader that I reserve the term "estimation" for the evaluation of a magnitude, whose exact value we do not know, but which does nevertheless exist, independently of ourselves. Thus it is only when I am assured, as in the present case, that the parameter has an objective meaning that I speak of *estimating* it. In the other cases, I use the verb "to choose." Let us always keep in mind the distinction between *objective statements, methodological choices and test criteria*: we *estimate* objective magnitudes, we *choose* methods and we *agree* on criteria.

Part II Criteria of Internal Objectivity in the Case of Unique Phenomena

"Omnis res positiva extra animam eo ipso est singularis." W. of Ockham

Approximate translation: "Every thing that exists outside the mind is, by that very fact, singular."

Chapter 3

The Poisson Forest

In the following pages and, except when otherwise specified, until the end of this work, we shall concentrate our attention on the case of a unique object or phenomenon – a forest, a mineral deposit, a mountain range etc... – which occupies a well-defined *bounded* part of that space in which we live. These objects are interesting in themselves. They can be, and actually are, studied from the panscopic viewpoint which is that of scientific enquiry. But they are also valuable and useful. The practitioner who is responsible for administering and exploiting them – the forester, the miner – meets problems of an essentially practical nature that he must solve, in one way or another, in order to fulfil his duties: for example, the miner must *estimate* the various parts of the deposit before he can decide which ones to mine and which ones to abandon because the grade is too low. In order to carry out this estimation, he has only at his disposal, by the very nature of things, very fragmentary and limited information. His viewpoint is therefore essentially monoscopic. Not that he has no interest in the other aspects of the object: on the contrary, one observes among professionals a sort of contained enthusiasm. But he must first of all think of fulfilling his task. In what follows, I shall sometimes adopt one viewpoint and sometimes the other, but the balance will be mostly in favor of *monoscopic models*. For it is in relation to the latter that the problem of objectivity arises in its most acute form. They are also the ones about which I have acquired the more direct experience through extensive practice. I can therefore hope that I shall know more or less what I am talking about when I speak of them.

External objectivity, that is, the objectivity of the general methodology we use to construct individual models, does not raise major problems: it is simply the sanction of practice. For example, Geostatistics can be considered as being objectively based because it has been successfully used to estimate several hundred mineral deposits. We shall therefore mainly deal, in what follows, with *internal objectivity:* that of statements concerning this particular deposit or that particular forest. We shall start with an introductory example, that of the "Poisson forest".

The Parameter θ: Does It Exist and is It Useful?

In order to get to the heart of the matter in a concrete way let us consider the Poisson model sometimes used by foresters to represent the distribution of trees (considered as points) over the forest (J. P. Marbeau[1]). Clearly, to say that the forest "is" Poisson is a condensed way of saying that the available data are not incompatible with this Poisson interpretation. According to the viewpoint of "orthodox" statistics, the most important problem one must then solve is that of *"statistical inference"*, that is, the "estimation" of the unknown *intensity* θ of the Poisson process. For once this parameter is known one can calculate all other characteristics of the process. This point of view attributes, implicitly, a real and objective existence to the intensity parameter θ: namely that even if our information allows us to arrive at no more than an approximate estimate of the "true" value of θ, this does not change the fact that the latter exists somewhere in nature, and could be precisely measured, given perfect information. But in reality it is not at all certain that this contention has an operational sense: for in order to determine θ precisely, the forest would have to extend to infinity (and remain Poisson) while its real size is in fact limited. The presumed obviousness of the existence of θ is based on a summary identification of the model (the Poisson process) with reality (the forest). Such a confusion, which is quite common among statisticians, is essentially an epistemological short-circuit. For however well a model is adapted to its object, we never have a guarantee that all its characteristics will faithfully mirror objective properties of reality which can be put in a one-to-one-correspondence with them.

If we firmly maintain this necessary distinction between the model and reality, the problem of statistical inference, (which is from the rigorous point of view insoluble) loses much of its importance. The parameter θ remains partly indeterminate, even if we have perfect knowledge of the forest. One may therefore seriously doubt its "objective existence", and therefore not take too seriously its indeterminacy. In fact, the practical problem which interests the forester is never the estimation of the Poisson intensity θ of the theoretical model. The latter, even when used, only intervenes as a convenient computational intermediary and always disappears from the final result. What the forester really wishes is, for example, to estimate the number N(S) of trees in a given area S, knowing the number N(s) of trees counted in the sampled area s. Or else (and this is the same thing) he will be interested in the mean number (S) = N(S)/S of trees per hectare in the area S. To estimate θ(S) and N(S) he will calculate the quantities:

$$\theta^*(S) = N(s)/s, \qquad N^*(S) = SN(s)/s \tag{1}$$

It is to be noted that the Poisson interpretation actually plays almost no part in the formation of these "estimators", which appear to be the most "natural" possible. In fact, these estimators turn out to be unbiased in the framework of a much less restrictive interpretation than the Poisson model. It is enough, for example, to postulate a certain type of stationarity (in the probabilistic sense of the word), i.e., a certain form of space homogeneity of the phenomenon, or a certain constancy of mean density. (The latter is a more vague formulation, but it relates to physical reality rather than to the model, and may be more meaningful to the practitioner.)

On the other hand, the Poisson character of the model forcefully intervenes when we calculate the *estimation variance* σ^2, i.e. the variance (in the model) of the error $N(S)-N^*(S)$. Assuming that the sampled area s is included in the area S to be estimated ($s \subset S$), it is not difficult to obtain the following expression:

$$\sigma_E^2 = \theta S(S-s)/s .$$

But in reality the paramter θ which appears in this formula is not known exactly. We therefore replace it by its estimator N(s)/s, which leads to the formula:

$$\sigma_E^2 = N(s)\frac{S}{s}\left(\frac{S}{s}-1\right) \tag{2}$$

in which no parameter of the hypothetical Poisson model appears any longer, but only experimentally known quantities.

So in the final outcome the model parameter θ appears nowhere, neither in expression (1) for the the estimators not in expression (2) for the variance we attribute to them. This parameter is thus never actually used, and since its objective meaning has already been put in doubt on other grounds, one starts to suspect that this famous problem of statistical inference is perhaps a false problem. On the other hand, however, in order to proceed from (1) to (2), i.e. in order to obtain the estimation variance, we use explicitly the Poisson "hypothesis", that is, the fact that the model we have chosen is of the Poisson type. Nevertheless, it is not necessary at any stage to *specify* the model, i.e. to fix the numerical value of θ.

To choose a *generic model* of Poisson type (independently of the value of the parameter θ which specifies it) simply means to formulate in probabilistic terms – with all the conceptual advantages and the operational possibilities this entails – an objective physical property which we could otherwise only describe in vague terms as:

(1) the "mean" density of trees per hectare does not show any systematic variation in space, and

(2) the fact that one region has a high (or low) density of trees does not imply "on the average" that the neighbouring region will have a higher (or lower) density than the others.

These two properties (even if it is difficult to formulate them precisely in non-probabilistic language) correspond undeniably to objective characteristics of the real phenomenon. They will certainly appear highly meaningful to the practitioner (the forester) who will not hesitate, in general, to state in each particular case whether they are plausible or not.

We see that *we are no longer rally dealing with the classical problem of statistical inference* (estimation of the intensity θ of a Poisson process) *but rather with the choice of an adequate model to represent a given physical reality*. The crucial point is that it is a certain physical interpretation of the reality (spatial homogeneity and "independence" of disjoint regions) which, once accepted, leads us to adopt the generic Poisson model. The problem of model specification (the estimation of θ) is now of limited importance, and can be even said not to arise at all: for neither the expression (1) for the estimator, nor the expression (2) for the variance we attribute to it depend on its value. It is only to simplify language that we shall say that *"this forest can be represented by a Poisson process with intensity θ = so many trees per hectare"* [and the numerical value attributed to θ will coincide with that of the estimator $\theta(S)$ of the mean density in S].

Three Steps in the Choice of a Model

The preceding analysis has highlighted three steps which have very different epistemological standings:

(1) there is first an *epistemological choice:* it has been decided to use probabilistic techniques to represent the phenomenon (the forest). This is a decision, not a hypothesis. It is a *constitutive decision.* (It "constitutes" the forest as an object of study, it defines the general framework within which we shall operate and determines the choice of the tools we use.) It is not an experimentally verifiable hypothesis (for it is neither true nor false to say that "this forest is a realization of a stochastic process", since there is no conceivable experiment or observation that could refute this proposition). At this level, we shall speak of a *constitutive model* (here a probabilistic one).

(2) We next encounter a *hypothesis about the physical nature* of the phenomenon studied – such as spatial homogeneity, absence of influence between neighbouring regions – which leads to the *choice of a generic model:* the process is a Poisson process. Contrary to the preceding choice (which can only be justified by its efficiency and by the successes it leads to, and on which one can only pass judgement in the long run, after having dealt with a large number of cases) this second choice follows from a physical hypothesis which can be objectively tested. It can, therefore, be supported or rejected by the experimental data either through statistical tests, which can be very easily devised for this very simple particular case, or through some other method, including the judgement of the practitioner who knows that forest well. Resorting to the practitioner's intuition has no mystical connotations whatever to it; it simply reflects the epistemological priority that we give to reality over the mathematical or rather statistical model we have chosen to describe it.

We cannot overstress the capital importance of this step. For it is essentially here that we incorporate in the model hypotheses that have an objective meaning and that carry with them positive information which is not contained in the raw data. *It is only because of this positive contribution that we can* (apparently) *extract from the data more than they really contain* (i.e. a prediction as well as an estimation variance). The counterpart of this small-scale miracle is that our model is now vulnerable, and that our predictions can now be contradicted by experiment if the hypotheses on which we base them are not objectively valid. It should be made clear here that it is not enough to check (e.g. through tests) that these hypotheses are compatible with the data, that is, that they are verified over the sampled plots. *We must also assume that they remain valid over the regions that have not been sampled,* and we shall only know this after the fact. The choice of the model constitutes therefore an *anticipatory hypothesis* and always introduces a *risk of radical error*. This is why it is imperative that it takes into account not only the numerical data, but also all other available sources of information (general knowledge about this type of phenomenon, the experience of practitioners, etc...).

In order to better localize the input of positive information and the consequent vulnerability of the model, it is often advantageous to distinguish two steps in the choice of the generic model: *choice of a generic model in the wide sense, then choice of a particular type* within this model. In the present case, the generic model (in the wide sense) would be for example "a stationary point process." The word "point" means that we have decided, as a first approximation, to treat each tree as a point in a plane. This is a (constitutive) decision that does not entail any input of positive information and therefore introduces no risk of experimental rejection. On the other hand, the word "stationary" is associated with a hypothesis of spatial homogeneity and entails an input of positive information as well as vulnerability, but to a relatively small extent. By *model type* we shall mean a model that needs only the numerical value of a small number of parameters in order to be completely specified. Here the model type chosen is a "Poisson process" with only one indeterminate parameter (the parameter θ). The choice of this type implies a very strong hypothesis, namely the lack of interaction between neighbouring regions. It introduces a large amount of information, which enables us, among other things, to calculate the estimation variance. And it therefore introduces further risks of experimental rejection. This is a general rule: very often it is *the choice of a type within the generic model that constitutes the crucial decision,* the one that opens the largest number of operational possibilities, but also inevitably introduces the greatest risk of error.

(3) Finally the last step is the choice of the specific model or, as we shall also say, the *specification* of the model (here the choice of the numerical value of $\theta =$ so many trees per hectare). While Mathematical Statistics attributes an absolutely vital role to this third aspect of "model choice," to which it refers as statistical inference (i.e. numerical estimation of the parameters), it will only play a very minor role, or even none at all, here since the essential results (estimation and variance) will be expressed, in the final analysis, in terms of the experimental data alone, in a form which exploits choices (1) and (2) (and particularly the choice of the Poisson type) but not at all choice (3) of the numerical value of θ. For in reality it is only for terminological convenience and to clarify our ideas that we carry out the specification of the model, while choice (1) (the epistemological choice) and choice (2) (choice of the generic model and its type) together

with the numerical information [here N(s)] provide us with an operational basis which is sufficient for solving the problem we are dealing with.

The precise distinction between these three aspects of model choice may appear elementary and uninteresting in a case as simple as that of the Poisson forest. But this will change rapidly as we examine more complicated models. Once the choice of the constitutive model (the epistemological decision) has been agreed upon, the crucial problem remains choice number (2), that of the generic model and more particularly that of model type. For once we have adopted a model that is suited to the experimental data, it is usually not difficult to specify it, if we really want to, on the basis of these same data. The important problem is therefore not that of statistical inference, but the choice of the generic model and of its type. Let us note carefully the fact that the problem is not just to "test a hypothesis" against the given data. This viewpoint, which is that of orthodox statistics, disregards the core of the problem. It is not only agreement with the available data that we need, but also *the much stronger hypothesis that the chosen type is also compatible with the data which are as yet unavailable, that is, the unknown or inaccessible parts of the phenomenon.* It is the latter hypothesis which explains the fertility of the method, and also, as we have seen, its vulnerability. Compatibility with the data is of course necessary, but it is never sufficient to insure us against an always possible disagreement with what is not given (and this is precisely what we are trying to estimate).

We shall study in Chap. 4, in a more general framework, the hierarchy of probabilistic models and the problem of choice as it presents itself at each of the three or four steps we have outlined. But first we shall critically examine our elementary example, which will enable us to clarify some basic guidelines.

Some Basic Guidelines

Let us see how things appear in the case of the Poisson forest. In order to study the objectivity of the model, we must clearly distinguish between two very different problems. We must *firstly* disregard the fact (which is in any case just a contingency) that only partial information is available and assume (without stepping out of the operational domain) that a complete topographic plot of the forest giving the position of each tree is available to us. Only concepts which are accessible on the basis of this maximal experimental basis will have an operational sense and an objective meaning. *Secondly,* we must examine the additional complications and uncertainties which are introduced by the incomplete character of the real information available to us. The problem is then: to what extent is an estimation carried out on that basis valid or even possible? The second problem is the most important one in practice, but from the point of view of methodology it is the first one that wins the day.

Assuming that the forest is completely known can we say in any objective sense that it is "Poisson," and, in particular, can we give an *operational* definition of the Poisson intensity θ? The answer is: yes, but only within certain limits. For example, if the forest has an area of, say, $S = 10{,}000$ hectares it can be divided into a number of adjoining plots of fixed area s. The area s should be neither too small (say at least 400 sq.m.) so that the representation of trees as points shall remain reasonable, not too large (say more than 100 hectares) so that a sufficiently large "statistical" sample shall be available. For each value of s we have a population of $n = S/s$ numerical values, namely the number of trees in each of the plots of area s, and we can check whether the corresponding histogram is, or is not, compatible with a Poisson interpretation. It is true that the answers provided by statistical tests are always equivocal. But since we are dealing with populations of at least 100 individuals, and in general many more (up to 250,000 for the smallest value of s), the answers we obtain will be sufficiently precise to allow us to neglect the margin of ambiguity or indeterminacy which is inherent, as we have seen, in every operational concept. If the populations are sufficiently Poisson, and if the corresponding experimental variances $\sigma^2(s)$ are approximately proportional to the areas s (and the acceptable degree of approximation can also, to a certain extent, be evaluated on the basis of statistical considerations) we can then say that we have exhibited the existence of an (approximate) law $\sigma^2(s) = \theta s$, which is precisely the *operational* definition of the "Poisson intensity" θ.

To this one must immediately add:

(i) that this intensity θ is not only measured, but also, in a deeper sense, *defined* up to a certain margin of indeterminacy that corresponds to the approximate character of the law $\sigma^2(s) = \theta s$ that defines θ.

(ii) Moreover, the concept of Poisson intensity loses all objective validity for areas s that are either too small or, more importantly, too large. For example, if $s = S/2$ (two plots) or $s = S$ (a unique plot) one can no longer speak of a Poisson population or of a variance that is proportional to s. In particular, to consider the total number N(S) of trees in the forest as a "Poisson variable of mean θS" does not retain any kind of objective meaning. We have gone completely outside the domain of validity of the operational concept.

Further remarks must be made. The Poisson model, even within the limits of its operational validity, is neither the only one possible nor even the best. Let us for the moment argue within a purely probabilistic framework. It is known that whatever the value of the parameter θ of the Poisson process, once the total number N(S) of points in the area S has been fixed, these N(S) points will be distributed in S independently of each other with a uniform probability density. In other words, conditionally on N(S) fixed, their distribution is independent of the value of θ. With N(S) fixed, the disjoint plots s are no longer completely independent [since the sum of the number of trees they contain must be equal to N(S)]. But for $S/s > 100$ their degree of dependence may be considered as negligible. The laws corresponding to this new model are no longer Poisson but binomial (although they differ but little from Poisson laws). Expectations and variances are given by

$$E[N(s)] = \frac{s}{S} N(S); \quad Var[N(s)] = \frac{s}{S}\left(1 - \frac{s}{S}\right) N(S) \tag{3}$$

Since s/S is $<1/100$, the variance is practically equal to (s/S)N(S); i.e. to the expectation. We obtain the same result as with a Poisson law with expectation $s\theta(S)$ (instead of $s\theta$), where $\theta(S) = N(S)/S$ is simply the classical "estimator" of the intensity θ.

For a statistician, this elimination of the parameter results from the fact that $\theta(S) = N(S)/S$ is a "sufficient statistic", that is, it contains all the information that is at our disposal to "estimate" θ, and this property expresses itself here by the fact that the "conditional" law of the variables N(s) associated with the different plots no longer depends on θ when N(S) is fixed. In this form it is a special property of the Poisson process, which would not be retained in the case of more general processes. But this way of presenting the matter amounts to a role reversal, since we are making a judgement about reality from the point of view of the model. I shall give a general formulation of the criterion of objectivity that we need later. For the moment let us continue with our study of the Poisson forest.

A Criterion of Objectivity: Regional Magnitudes

At this stage of the analysis we begin to suspect that the parameter θ of the theoretical model may well be devoid of objective existence, since the objective characteristics we had attributed to it, including its operational definition, can be taken over equally well by the so-called "estimator" $\theta(S) = N(S)/S$ [in fact better, since formula (3) for the variance retains a meaning for $s = S$. The variance of N(S) is simply zero, since N(S) is fixed]. We see that the viewpoint of orthodox statistics is now reversed: what was considered a reality, unknown but having an objective existence, and which was the object of a "statistical inference" (the parameter θ) turns out to be a useless fiction and it is the so-called estimator [the mean density $\theta(S) = N(S)/S$ of the real forest) that alone has now an objective meaning.

Let us agree to call *regional magnitude* or more simply *regional* any magnitude that, like N(S) and also all the numbers N(s) associated with all possible plots s, is unequivocally defined when the regionalized phenomenon in which we are interested (here the forest) is sufficiently known in the totality of its spatial extension (or field) S. The Poisson parameter θ is not a regional magnitude and is also devoid of objective meaning. We shall now state a first principle, which is useful for proper orientation in the choice of probabilistic models to describe *unique phenomena:*

Principles of regional magnitudes. Among the parameters of a probabilistic model that describes a unique phenomenon *only those that can be expressed in terms of regional magnitudes have an objective meaning*. Similarly, among the concepts that are introduced by the model, the only ones that have an objective meaning are those that can be given an operational definition in terms of regional magnitudes; that is, a definition constituted by a system of experimentally verifiable relations (within a given domain of validity and with a given degree of approximation) between these magnitudes.

Thus the concept of distribution (Poisson or binomial) of the number N(s) of trees in plots of size s has undeniably an operational meaning as long as s is neither too large nor too small. On the other hand, for S = s, N(S) as a Poisson variable with expectation θS has no operational meaning.

From the Vantage Point of the Practitioner

Let us now turn to the second problem, that of the possibility of an estimation based on partial information. In particular, we must ask ourselves whether the concept of *estimation variance* introduced previously can be given an operational definition in terms of regionals. The answer clearly depends on whether we stand either *"after the fact"* (that is, if we assume that the forest is perfectly known, and we then compare with reality the estimation we have made on the basis of some partial information that we have chosen at will) or *"in the heat of action"* (that is, if we assume that we really have only partial information at our disposal).

After the fact, the answer is certainly positive: we can define the estimation variance in terms of regional magnitudes. (In the general case, there are even several non-equivalent ways of doing this, and we shall devote our attention later to this difficulty. But in the Poisson case all these definitions are equivalent, and the ambiguity disappears). In fact, formula (3), which gives the variance of N(s) for N(S) fixed, provides us with an experimentally verifiable relationship between regional magnitudes: to the extent that it has been actually verified, we are entitled to deduce from it the variance of the error associated with the estimation of N(S) on the basis of the information N(s) supplied by the sampling of an area (of arbitrary shape) of size s, giving

$$\sigma_E^2 = \left(\frac{S}{s} - 1\right) N(S) \tag{4}$$

If we now situate ourselves "in the heat of action," N(S) becomes unknown (since it is precisely the magnitude we seek to estimate) and all we can do is to replace it by its estimator (S/s)N(s). We are now back to formula (2), which is a reasonable approximation to (4), provided s is not too small.

The objective value of formula (2) – the only one we can use in practice – is therefore inferior, but nevertheless comparable, to the operational definition (4). It is just that the validity of the latter can only be finally established after the fact, when the whole forest is known, and not on the basis of the partial information available "in the heat of action". Therefore, also, the value of (2) depends on the validity of an *anticipatory hypothesis* (which may turn out to be false after the fact) according to which the Poisson character of the forest, which we have successfully tested on the part s that has been sampled, may be extended to the total area S.

This is a general situation: we are never protectd from nasty surprises. The fact that a model fits a partial sample well never guarantees that it will remain compatible with the totality of the phenomenon. *Consequently, any prediction* (including that of an estimation variance on any other concept that can be expressed in terms of regional magnitudes) made on the basis of partial information and a model *may turn out to be grossly in error*. The degree of confidence we can attribute to it depends on the objective value of an anticipatory hypothesis that we shall only be able to verify after the fact. In scientific matters, one can always suspend one's judgement until sufficient information is available. The practitioner, however, has no option but to take a decision in the heat of action, that is, on the basis of information that is in general insufficient, and this always implies a *risk of radical error*.

Note

[1] "Géostatistique forestière", Thesis, Fontainebleau 1976.

Chapter 4

Choice and Hierarchy of Models

Let us now define more precisely the various levels that we have met in the simple example of the "Poisson" forest: the constitutive model, the generic model, its type and its specification. The situation we are referring to is as follows: we are interested in a certain phenomenon or a certain object that is defined on a *bounded* domain S of Euclidian space, with 1, 2 or 3 dimensions (possibly 4 if we also wish to include the time dimension). Let us assume that this phenomenon may be satisfactorily described by a function z (or perhaps several) defined on S. We shall call such a function, in general terms, a *regionalized variable* (REV). The expression *"after the fact"* will be taken to mean throughout that we mentally put ourselves in the ideal situation where we know the numerical value of z(x) at every point $x \in S$. On the other hand, the expression *"in praxi"*[1] will refer to the situation which is usually met in practice, where we only have at our disposal a sample of the REV, for example the N numerical values $z_\alpha = z(x_\alpha)$ taken by z(x) at the N given experimental points x_α, $\alpha = 1,2,\ldots N$. (One may also, of course, consider the case where the z_α are the numerical values of some functionals of z, e.g. the mean values of the REV over given sets $s_\alpha \subset S$, or values that are distorted by measurement errors etc...). But it is understood that even "in praxi" we have, over and above this numerical information, a certain knowledge, of structural and qualitative nature, about the physics of the phenomenon, and certain ideas (either intuitive or rigorous, but which may eventually prove erroneous) about the behaviour it is susceptible of showing. We may also have acquired some experience from having dealt with similar cases, advice (possibly contradictory) from various specialists and practitioners etc... These various sources of information, in spite of their heterogeneous nature and their unequal quality, must be given much weight when we choose the model, that is, when we adopt, "in praxi" an anticipatory hypothesis. We shall call them, in general terms, *nonnumerical, or qualitative information.*

A First Criterion: Decidability Through Regionals

We shall use the term *regional magnitude* or simply *regional* to describe any functional of the regionalized variable z defined on S, that is, any magnitude whose value is determined when all the numerical values of z(x), when x ranges over the whole of S, are given. For the moment at least *we shall use the regionals* as a criterion of objectivity. Within that framework (which will turn out to be too strict) *a statement will have an objective value if, and only if, it is decidable after the fact:* that is, if it can be declared unambiguously to be true or false once we know all the values z(x), $x \in S$. We may provisionally use decidability and not falsifiability alone, as suggested by Popper's criterion because, after the fact, by definition, all the "virtual falsifiers" of the statement have themselves been decided, that is, uniquely classified as true or false. In any case, under these circumstances, the *"truth value"* of the statement (1 if it is true and 0 if it is false) is itself a functional of z, and *is therefore itself a regional magnitude.*

Later on we shall make this too rigid criterion more flexible, and return to the Popperian criterion of falsifiability alone. For the regionalized variable z is not in any way identical with reality. It is itself a first model. In nature, there are no such things as points $x \in S$ or numerical values z(x): *we* are the ones who have introduced them as the product of a preliminary pruning or simplification of the overabundance of reality. This departure from reality that we have introduced at the outset remains unperceived most of the time but will reappear on certain critical occasions, when the necessity for a finer analysis of the physical meaning of certain mathematical concepts (such as continuity or differentiability) will be felt. We shall return to this difficulty in the section devoted to the *primary model,* and we shall argue, for the moment, as if we had the right to identify the REV with reality.

According to this criterion, statements such as: "the regionalized variable z is a realization of a random function" or even, more precisely, "of a stationary random function" *have no* objective meaning. They could not be found to be true or false even if we knew all the numerical values of z, that is, if we had all the possible empirical information that we can acquire. Similarly, a statement such as: "there are 99 chances out of 100 that the mean value of z in S will be larger than 2.5" has no objective meaning, at least according to the present criterion, which is a criterion of internal objectivity. In practice, however, one does make such statements, and one is right to do so. One is even forced to do so; but then a different criterion, which we have called external or methodological objectivy, must be invoked.

Among the parameters that specify our model, only those that we can identify with regional magnitudes will have an objective meaning. We shall call them *objective parameters* and the others will be called *auxiliary or conventional parameters.* For example, in a model where the REV z is considered as a realization of a stationary RF (random function) with expectation $m = E[Z(x)]$, this expectation *is not* an objective parameter, but a conventional one. Similarly, even if a concept has a mathematical definition in the model, it will only be considered *operational* (within a well-defined domain of validity) if it is possible to redefine it by means of a system of experimental relations between regional magnitudes that is testable after the fact (with a given approximation and within a given domain).

The Constitutive Model

In the problem at hand, the "constitutive" model is that of a random function (RF). To choose this model means that we have *decided* to consider the REV z(x), $x \in S$, as the restriction to S of a realization of some RF Z(x) that we have of course to specify later on. But at this stage we are just dealing with the (mathematical) concept of RF in general. We have simply made a methodological choice, taken a decision; and this is not a verifiable or falsifiable hypothesis. We have already seen that the proposition "the REV z is a realization of a RF" is devoid of any objective meaning. This does not mean that it is an arbitrary decision. If we have taken it, it is because we have good reasons to believe that it is the best decision possible, perhaps the only one that will supply us with a conceptual framework within which we hope to formulate and solve in the best possible way the problem or problems that occupy us.

Before taking it we certainly have tried, in this case or in similar ones, to fit a deterministic or functional model which would of course allow us to make more precise predictions, and have failed. (All this is, of course, in praxi: after the fact, we have a very precise functional model, namely the function z itself. But even after the fact this perfect model may turn out to be far too complicated for convenient use.)

This choice is not constitutive as far as the phenomenon itself is concerned (for it exists and is what it is independently of the decisions that we take about it), or as far as the REV z(x) is concerned. For the latter is itself a first constitutive model that precedes the RF: it defines the scope, and is itself the subject of, a preliminary domain of study, and thereby achieves a first reduction or simplification of the overabundance of reality. But the choice of the model "RF" is constitutive insofar as it creates or constitutes the object that we shall actually study, namely the REV considered as a realization of a random function. It also determines in advance the choice of the methods that we shall use in our study: probabilistic techniques.

The intuition that has guided this fundamental choice is clearly that of the coexistence, within the same phenomenon, of a structured aspect and of a "random" or chaotic one. We suspect the existence of certain structural traits that are masked by the chaotic variability exhibited by the phenomenon at the local level, excluding the possibility of any simple functional representation.

This probabilistic intuition is relatively *fleeting* and difficult to nail down. It may even disappear completely with a different choice of constitutive model.

For example we could choose to represent the REV z(x) by an expansion of the type

$$z(x) = \Sigma a_n f_n(x) \quad (x \in S)$$

where the f_n are functions "orthogonal" on S (e.g. trigonometric functions if we wish to carry out a harmonic analysis of the phenomenon, or orthogonal polynomials etc...). Once the f_n are chosen, the probabilistic interpretation consists in further interpreting the coefficients a_n as realizations of random variables A_n. But this interpretation reeks of artificiality. For if we know everything that can be known about the REV z(x), namely the numerical values of the coefficients a_n, all that we can say about each random variable A_n is that a unique draw has attributed to it the numerical value a_n. From the classical point of view, we are in a situation where statistical inference is completely indeterminate (since one cannot reconstruct the law of the random variable A_n on the basis of the unique numerical value a_n). In the present terminology we shall say that the laws of the A_n do not correspond to any operational concepts. The model of random A_n is not specifiable in terms of regional magnitudes.

In reality we have no need to probabilize the coefficients a_n of the expansion of z(x). Once the REV is completely known these coefficients are numerically determined and are therefore objective parameters. If our aim is to perform a spectral analysis of the energy content then the $|a_n|^2$ provide us with all that we need. If we are looking for a simplified, smoothed, version of the phenomenon, we can split z(x) into two terms:

$$D = \sum_{n=0}^{N} a_n f_n ; \quad R = \sum_{n=N+1}^{\infty} a_n f_n$$

the trend and the residuals. (The choice of N is clearly arbitrary, but in practice, if we keep in mind the aim pursued, there is very little ambiguity.) To say at this stage that D represents the deterministic part of the phenomenon and R the random part is a rather empty manner of speech, since the two expressions are of the same nature and have exactly the same epistemological status. Probabilistic intuition has well and truly been cleared out of the picture.

In praxi some difficulties arise. For on the basis of limited information (e.g. some experimental points) we may be able to evaluate approximately the first few coefficients a_n, e.g. those that appear in the expression for D. But in the absence of a probabilistic model we shall find it far more difficult to put some bounds on the order of magnitude of the error committed in this estimation.

The Generic Model: Type and Specification

The main step is now to choose, among all possible RFs, that which is to represent the phenomenon we are interested in. We shall denote this random function by Z or Z(x), $x \in S$. To make our choice we have at our disposal on the one hand qualitative information about the phenomenon, and on the other hand the REV z(x), $x \in S$ (which is only partially known in praxi), considered as a realization of Z. It is at this point that we run into the difficult problem that classical terminology calls statistically inference. For a random function Z is usually defined by its *spatial law*, that is, by all distributions (which are infinite in number) of the form

$$P[Z(x_1) \leqq z_1; Z(x_2) \leqq z_2; \ldots Z(x_n) \leqq z_n]$$

for all integers $n \geqq 1$ and all possible choices of the support points $x_1, \ldots, x_n \in S$ and the real numbers $z_1, \ldots, z_n$. The problem is that it is not possible to carry out an estimation of all these laws on the basis of the information that is available in praxi, which consists of a finite number of samples drawn from S. In contrast with the situation in physics, for example, where one has in principle the possibility of repeating the same experiment as many times as one wishes, and where therefore one may have several realizations of the same RF, we are dealing here with a unique realization, sampled at a finite number of points only. Statistical inference of the spatial law (in its totality) is certainly not possible "in praxi". Moreover, and this is even more important, it is easy to see that it is not even possible "after the fact", that is, when we know z(x) for every $x \in S$. From the classical point of view, we have run into an insoluble or indeterminate problem. But as far as we are concerned, keeping in mind our criterion of objectivity in terms of regional magnitudes, we must rather conclude that this is a false problem: *the spatial law is not uniquely defined by the REV, and is therefore devoid of any objective meaning.*

Consequently we must, in one way or another, introduce some *restrictive conditions* in order to lift the indeterminacy. From the classical point of view these restrictive conditions are most of the time considered as "hypotheses". (For example the "hypothesis" of stationarity) which, *if they are verified,* will allow us to reduce the number of parameters to be estimated and will make statistical inference reasonably feasible in praxi. The problem is, are they verifiable after the fact? For even from the classical point of view a hypothesis that is too general, like stationarity, cannot be the subject of meaningful statistical tests. Indeed, any function z(x) defined over a bounded domain S can always be considered as the restriction to S of a periodic function whose periods along the coordinate axes define a parallelepiped which is large relative to the domain S. If we now translate x by a random vector uniformly distributed over the parallelepiped of periods, then this periodic function becomes, in all rigour, a stationary random function. And the given function z on S can, in all reason, be considered as a realization of it. It follows that no test of "stationarity" in general will ever lead to a negative answer.

This conclusion should not surprise us: it simply means that "stationarity in general" is not decidable [2] in terms of regional magnitudes, and has therefore no objective meaning. Consequently, among the restrictive conditions that we shall introduce, we shall have to distinguish carefully between those that really correspond to *objective hypotheses* (decidable [3] after the fact) and those that are only *methodological choices*. Only the former will bring in substantial supplementary information that will enable us to extract from the numerical data available in praxi more than they actually contain. *They* are the ones that constitute, properly speaking, the anticipatory hypothesis, already mentioned several times, which may be found false after the fact. It is this vulnerability that is the necessary counterpart of their fertility. In order to reduce as much as possible this vulnerability we must choose them in praxi according to a principle of strict economy. From the monoscopic perspective that is ours here *an objective hypothesis should be adopted in praxi only if it is absolutely indispensable to attain the proposed goal* (and naturally only if it is not incompatible with the numerical and qualitative information at our disposal). The latter, on the other hand, since they introduce neither positive information nor any risk of vulnerability, are simply methodological *decisions*. As far as they are concerned, the rule will thus be to choose those that lead to the most efficient solution of the problem at hand. Let us also note, however, that each time there are methodological choices to be made, the *manner* of making the choice in each particular case depends on a more general methodology that will be submitted in the course of time to the sanction of practice, and is therefore not devoid of objectivity in the long run, even though in each particular case they remain in some sense arbitrary.

At any rate, once these restrictive conditions (objective hypotheses and methodological choices) have been introduced, we do not need look for our RF Z(x) among the infinitely too vast set of all possible RFs but only among the far smaller family $\mathscr{F}$ defined by these conditions, and within which the number of parameters that define an individual RF is substantially reduced (even though it may still remain infinite). Naturally, if we introduce additional restrictive conditions, we shall obtain a hierarchy of models and submodels $\mathscr{F}_0 \supset \mathscr{F}_1 \supset \ldots$ ranging from the most general to the most particular. In that case we say that $\mathscr{F}_0$ is a *generic model* and that a submodel $\mathscr{F}_1 \subset \mathscr{F}_0$ that is particularized to the extent that in the identification of each individual model in $\mathscr{F}_1$ only a *small number of parameters* (two ot three for example, rarely more) intervene is *a type* of the generic model $\mathscr{F}_0$.

For example the family of stationary Gaussian random functions constitutes a generic model $\mathscr{F}_0$. In order to define a random function $Z \in \mathscr{F}_0$ an expectation m and a covariance *function* $\sigma(h)$ must be provided. The submodel $\mathscr{F}_1$ of stationary Gaussian RFs with exponential covariance $\sigma(h) = \sigma^2 \exp(-a|h|)$ constitutes a submodel or type of RF in $\mathscr{F}_0$ within which an individual will be identified by three parameters (the expectation m, the variance σ^2 and the scale parameter a of the covariance).

Instead of RFs, we very often consider *equivalence classes* of RFs and we also call a family $\mathscr{F}$ of such equivalence classes a generic model. For example, in problems of *linear* estimation, it is only necessary to know the moments of order 1 and 2 of the RF. We call the class of all RFs with a given expectation m(x) and a given covariance $\sigma(x, y)$ a random function of order 2. Thus the family of stationary RFs of order 2 is a generic model $\mathscr{F}_0$ where each individual is specified by an expectation m and a stationary covariance function $\sigma(h)$. The stationary RFs of order 2 with exponential covariance then form a type $\mathscr{F}_1$ in $\mathscr{F}_0$, where each individual is defined by 3 parameters only (m, σ^2 and a). But each such "individual" is in reality itself a class of RFs (that we do not distinguish from each other). We say that it constitutes a *specific model* or a *specification* of the generic model $\mathscr{F}_0$ (and of the type $\mathscr{F}_1$).

This three-level hierarchy (generic model, type and specification) may appear arbitrary. It proves however to be useful in practice because very often the choice of a generic model is a simple methodological decision, while the choice of type almost always has an objective import. *It is generally when we choose (in praxi) the type within the generic model that we introduce the most vulnerable (and the most fertile) anticipatory hypothesis.* The specification (in praxi) or in other words the *estimation* of the few parameters on which the type depends, on the basis of the available information, obviously entails also serious risks of error. (This is the classical problem of "statistical inference".) But the nature of these errors is perhaps less fundamental. For once we have chosen the type of model, mathematical satistics (applied within the framework of the given type) may indicate to us whether the estimation of the parameters can be carried out in a reasonable way or not on the basis of the available information. If it cannot, this means that we cannot solve the problem that interests us within the framework of the chosen generic model and type and we must look for another formulation. Let us also note that, for reasons that are easy to understand, the qualitative (non-numerical) information intervenes in an essential manner in the choice of type while type specification is most frequently carried out on the basis of the numerical information alone.

In summary, a generic model is characterized by its *extension and its specification*. The extension is simply the set of all RFs that belong to the model. For example, in the generic model "stationary RF of order 2", the extension is simply made up of all RFs that have stationary moments of order 1 and 2. On the other hand, the specification defines (or specifies) the classes of the generic model, also called specific models. In our example, a specific model is made up of all the RFs that have the same expectation and the same covariance, so that the specification consists in supplying a constant m and a positive definite function $\sigma(h)$. Midway between these two concepts we meet the concept of *type* or submodel where the specification reduces to that of a small number of parameters (for example: stationary RFs of order 2 with exponential covariance).

As far as specification is concerned, it is important to know whether it has an objective meaning or not. From that point of view the previous example is rather badly chosen, for neither m nor $\sigma(h)$ are strictly identifiable with regional magnitudes. Similarly the parameters m, σ^2 and a that appear in the exponential covariance type are *not* strictly objective parameters (except perhaps if the domain S is very large compared with 1/a). On the other hand we shall see that the product a σ^2 (that represents the slope of the variogram at the origin) has an objective meaning.

From the methodological point of view, let us also note that to adopt a given generic model implies that *we give up the possibility of distinguishing between different RFs that belong to the same specific model,* that is, having the same specification. As an immediate consequence, *one may not subsequently use tools other than those contained in the specification itself.* For example, in the case of the generic model "stationary RF of order 2", the only allowable tools are the expectation m and the covariance $\sigma(h)$. This means, for example, that as far as the estimation problem is concerned, we restrict ourselves to the class of affine linear estimators [of the form $Z^\star = \lambda_0\delta + \Sigma\lambda_i Z(x_i)$], and, from among all the probabilistic characteristics of estimators of that form, use only their means and variances, which alone can be expressed in terms of m and $\sigma(h)$. To be completely rigorous, it would be necessary to be even more severe, and to limit oneself, among estimators of that type, to those whose characteristics (expectation and variance) may be redefined after the fact in terms of regional magnitudes. In practice, one often compromizes over this rule, which is rather too strict, but in general it is advisable to try to conform more or less to it.

Preliminary Remarks on the Criteria of Choice

It is not possible here to examine in depth the rules and precautions that one must observe when choosing (in praxi!) a model, and we shall just give a few brief guidelines. It should also be understood that we are resolutely adopting the monoscopic point of view, which is what the practitioner does when he must make a decision. In a more disinterested type of research, one would rather aim at a model that would be as panscopic as possible, and would therefore present *as high a degree of specification as possible on the basis of the available information.* For the more detailed the specification of the model is the richer it becomes and the better able it is to represent the various aspects of reality. The practitioner, on the other hand, must weaken the anticipatory hypotheses on which he must perforce rely as much as possible, in order to best insure himself against the risk of making a wrong decision. He will therefore look for *the weakest possible degree of specification compatible with the effective solution of the problem at hand.*

In all cases the model must satisfy the following conditions:

(a) It must be *operational:* once *specified,* the model must allow us to obtain an effective numerical solution of the problem. It follows (and this is consistent with our monoscopic viewpoint) *that the definition of the model to be used is closely related to the nature of the aim pursued.* This is a fundamental methodological point, to which we shall return. *The economy principle* that presides over the choice of monoscopic models may be expressed as follows: *the specification must contain all that is needed to solve the given problem, but it is unnecessary that it should contain any more than what is needed.*

(b) It must be *specifiable*. In vague terms, which we shall try to make more precise later on, this means that it must be "reasonably" possible, on the basis of the (numerical and quantitative) information available, to *choose* the type of model and to *estimate* the numerical values of the parameters that specify it.

(c) It must be *compatible* with the (numerical and quantitative) data in a sense that we shall progressively make more and more precise. The problems raised by this condition and the previous one are analogous (but not identical) to those that classical statistics knows under the name of "test of hypothesis" and "statistical inference" of the parameters. They have only partial relevance to mathematical statistics. For on the one hand, if a parameter is devoid of objective meaning, it should not influence significantly the solution of a real problem. (For a real problem must be formulated and solved in operational terms, i.e. in terms of regional magnitudes.) This parameter may therefore be chosen in a relatively arbitrary fashion. The fact that statistical inference may not be possible loses much of its importance. Conversely, if a "test of hypothesis" supports the acceptance of the type of model chosen, that is, it is found to be compatible with the actual data, this does not mean that a different kind of test would not have rejected the model. And above all nothing guarantees us that future data, those that are not yet at our disposal, will confirm our conclusion. The risk of radical error is always present. Finally, we must admit in all truth that it is rare in practice that really sophisticated tests can be carried out: either simply because none exist, or else, if they do exist, because they are too complicated, or because carrying them out requires information that is not available. (For example, to test moments of order two, it is necessary to know moments of order four etc...)

(d) It must be *efficient,* that is, it must lead to a correct solution (at least in a statistical sense) of the given problem. This point can of course only be verified after the fact. But the experience acquired with similar cases also plays an important role: the sanction of practice, similar to natural selection, mercilessly eliminates inefficient methodologies and inadequate models in the long run.

As far as point (a) is concerned, let us note the decisive importance of the *choice of the method* adopted to solve the given problem. The problem is what it is: it is a constraint. But in order to solve it we usually have a choice of several methods, some more and some less powerful, some more and some less approximate, some requiring more and some fewer prerequisites for their application. Our first task is to carry out, for each possible method, a strict analysis of its actual *minimal prerequisites*. (This pruning operation often reveals that certain parameters, which were thought to be indispensable, are in fact completely superfluous.)

Since it is in our interest to have a model that is as little specified as possible, we try, as far as possible, to make the specification of the model coincide with the minimal prerequisites required by the chosen method. This is one of the guiding rules in the choice of a generic model. The specification must contain the minimal prerequisites, but it is unnecessary that it should contain anything else.

The more powerful a method, the more minimal prerequisites it requires, and therefore the more the model must be specified. But there is always a limit, since the available data do not allow us to go beyond a certain degree of specification. This brings us to the second rule, which relates this time to the choice of the method that is to be used in solving a given problem: use the most powerful method among those whose minimal prerequisites can be reasonably specified on the basis of the available data.

The third rule is, obviously, that the model should not be incompatible with the experimental data or with our physical intuition of the phenomenon. Its compatibility with reality itself, however, is never absolutely assured, even if a critical examination of the possibility of specification (i.e. of the estimation of the regional magnitudes required) yields a positive answer. One is never completeley safe from nasty surprises. But there is no choice: one must, at some stage, introduce a hypothesis that enables us to connect, in one way or another, the structure of the unknown part of the phenomenon with that of the available data.

The Primary Model

In the preceding pages, we have argued as if it were possible to identify reality with the bare set of numerical values of the regionalized variable. But the latter never exhaust the abundance of reality. There is always something else, apart from punctual numerical values. Certainly we can define a mountain range as a collection of topographic level curves, but this analytical and exhaustive point of view may not necessarily be the most interesting. There is also the global geological structure of the range, its history, its tectonics etc... To view the whole picture, the bare topographical map is not sufficient, and we must introduce as many REVs as there are geological magnitudes to consider. We must also continually keep in mind the evolutionary aspect of the phenomenon and its genesis, even if direct observation does not provide us with anything in that direction. In one sense, therefore, reality is inexhaustible. But, nevertheless, the modest technical problems that we are considering, and which, by their very nature, centre on estimation and interpolation, fare rather well under a reduction in the abundance of reality to one or more functions z on S (the REVs). The remainder – the ideas we have formed about the genesis and the structure of the phenomenon and more generally the physical intuition that we have developed – will, however, continue to play a most important role behind the scenes. In general it is in that archetypal treasure house that we have the opportunity to find the schemata and driving principles of really well-adapted models. We should therefore be wary of the purism that attempts to identify reality with a set of numerical values. Firstly because numerical values are not reality but only a preliminary picture of it, that is analytically very rich but structurally very poor.

Secondly because the schemata of physical intuition supply us with the very guidelines that we need to orient ourselves in the problem of model choice. Nevertheless it is the analytical reduction of reality to a collection of numerical values that alone furnishes us with the basis of an operational methodology. Among the different possible models, and the various concepts that one may relate to them, we shall retain as objective (non-methaphysical) only those that can be made completely explicit in terms of regional magnitudes (i.e.: determined by the knowledge of one or more REVs over the domain S).

The situation here is somewhat as in logic: one knows well that in a concept there is much more than its extension. There is the driving element, the dialectical movement etc... But a rigorous formalized treatment is only possible if one agrees to identify the concept with its extension. This is the price we have to pay to avoid metaphysical illusions and to obtain an operational formalism. The rich content of the concept, fascinating and beyond grasp as it is, has apparently vanished, leaving behind a rigorous, operational, cold and somewhat boring methodology. But in reality nothing has been lost. For the driving element of the concept secretly continues to inspire our endeavours (and without it we would probably not attempt anything). However, these endeavours and their results must be immediately translated into the language of the most refined and rigorous formalism, which has alone won recognition.

The regionalized variable is thus not identical with reality, but constitutes itself a *primary model*. This is why the criterion of simple decidability in terms of regionals proves to be insufficient as soon as the need for a finer analysis is felt. One must then return to Popper's criterion, based on falsifiability alone.

This necessity manifests itself in a commanding manner every time one must take into acount the physical significance of the magnitudes that we are manipulating and the bounds outside which the concepts that correspond to them cease to be operational. For in nature there are neither points $x \in S$ nor numerical values $z(x)$ uniquely allocated to each point x. For a physicist there are no points in the mathematical sense but only elements of volume that are small relative to the scale that we have chosen to work at but large relative to the lower scales at which the phenomenon that we are studying ceases to be definable in operational terms. If, for example, $z(x)$ is supposed to represent the grade of some metal at a given "point" x in a mineral deposit, the reference volume element δv will usually be chosen large with respect to granulometric dimensions. If it is desired to study the phenomenon at the latter scale, which is possible, the grade $z(x)$ will appear as "all or nothing." It will be almost zero in the waste grains, and almost constant in the mineralized grains. In all cases its value will depend on the mineralogical constitution of the mineral. But even at the granulometric scale the elements of volume δv, although small in relation to the grains, remain large in relation to atomic or infra-atomic dimensions. Anyway, at the latter scale, we could only observe a chaos of wave functions. The usual model of three-dimensional Euclidian space would cease to be operational and there would no longer be any question of either points $x \in S$ or uniquely defined grade $z(x)$.

Thus the concept of REV, like any other physical concept, is subject to rather strict limitations outside of which it ceases to be operational. Keeping this in mind, we must not take literally purely mathematical statements concerning, for example the continuity or differentiability of REVs, or more generally any topological property that involves all neighbourhoods, however small, of a given point x. These are always characteristics of the model only, and not, in any rigorous sense, properties of reality, since below some threshold of smallness the phenomenon either, so to speak, vanishes or its nature changes.

Nevertheless, the contrast between these various scales is such that we may, as a first approximation, neglect these fuzziness effects (which will not manifest themselves at our scale of operations) and take the model "REV" as truly representative of reality. But even then serious difficulties remain. For the set of numerical values $[z(x), x \in S]$ is *infinite* and is therefore not amenable to exhaustive experimental knowledge. Thus a statement like "z is continuous (or differentiable) at the point x" that involves an infinity of points and numerical values (perhaps even a non-denumerable one) is, rigorously speaking, beyond any possibility of effective experimental control, and must therefore be considered as lacking any objective meaning. In fact, whatever the "real" function z is (in as much as it is definable), we can always find a continuous, differentiable etc... function Φ that is near enough to it to be experimentally indistinguishable from it. For we shall never know the $z(x_i)$ except over a mesh $\{x_i\}$, as tight as we like, but nevertheless always finite, and it will always be possible to interpolate them with a function Φ as regular as we wish.

Obviously, from the physical point of view the key notion here is that of *scale*. The function Φ chosen as above may be extremely regular (differentiable etc... in the mathematical sense) but only at the scale of the extremely tight mesh defined by the $\{x_i\}$. This very high microregularity does not preclude the function from behaving, in practice and at a higher scale, like an extremely irregular function, even like white noise.

If one studies the phenomenon at the latter scale, it would be physically *wrong* (although mathematically correct) to consider the function Φ as continuous and differentiable.

Thus the most usual mathematical concepts, such as continuity, differentiability etc... must be the object of an operational reconstruction. To treat a prob-

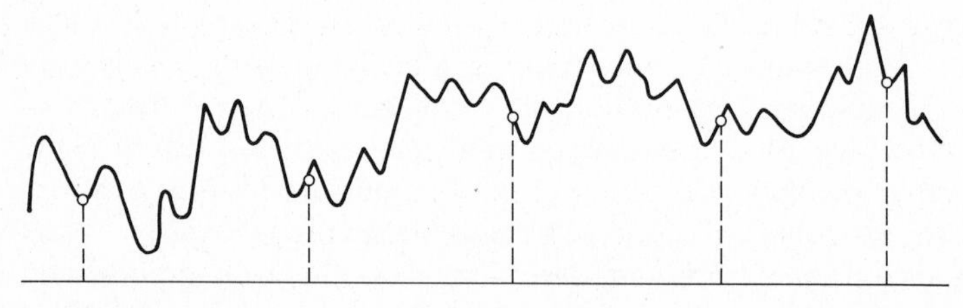

Fig. 1. It is mathematically correct but physically wrong to consider this function as differentiable

lem that is stated (for example) at the scale of one metre or of one hundred metres, it is enough to know z(x) over a mesh of one centimetre, or better to know over that mesh the smoothed version of z(x) over a one centimetre support. For the physical data correspond to little integrals rather than to strictly punctual values. One can therefore, in principle, have a physically exhaustive definition of the REV by means of some number (which may be large) of numerical values that are experimentally measurable. It is at that scale (centimetric in our example) that we must carry out the operational redefinition of the mathematical concepts.

Threshold of Robustness and Threshold of Realism

Let us further note an imortant aspect of the REV and of the model: namely that the model, once chosen, tends to live its own life, which partially eludes us, and coincides only in part with our physical intuition of reality and with the characteristics of the REV that we have chosen. This is because the model tends to go beyond reality, and if we let ourselves be carried away by mathematical formalism without critically questioning its relevance, if we yield to the temptation of *passing to the limit,* we shall be drawn beyond the domain where the model is operational. In other words, some results drawn from the model by simple mathematical deduction correspond to real properties and some do not. Let us note that this is so even when the model has been specified *after the fact,* on the basis of all the numerical values of the REV. This *threshold of objectivity* beyond which one is not justified to carry mathematical deduction, lest one produce artificial and false problems, is sometimes difficult to define precisely. But it is clear that it always exists and that the physical intuition of the researcher plays an important role in its delineation in practice. The situation is even more serious *in praxi,* because we may be induced to err in adopting objective anticipatory hypotheses that may prove to be false after the fact. There thus appears in praxi a *threshold of robustness* that one should not overstep, or perhaps two thresholds, since we can distinguish a *robustness of type* and a *robustness of specification*, which refer to the possibility of error in the choice of type and in the estimation of the parameters that specify it respectively. As a general rule, an operation can be said to be "robust" with respect to a cause of error if its result is only weakly affected by this cause of error. Specific robustness may be studied quantitatively, since one may always vary the parameters of the model around the chosen values, and study to what extent these variations affect the results of the given operation. Robustness of type, on the other hand, is much more difficult to study, since it is related to uncertainty about a structural choice, and there usually is an unlimited number of types that can be imagined, all approximately equally compatible with the available data and information. So common sense, physical intuition and acquired experience should have their say here as well.

As far as these thresholds are concerned, what is fundamentally important is to always keep clearly in mind *the distinction between the characteristics of the model and the physical properties of the real phenomenon*. Within the framework of probabilistic interpretation, this amounts, in broad terms, to distinguishing carefully between *the RF itself* (a mathematical concept), as defined by its spatial law, or simply characterized, in practice, by a covariance function or a variogram, on the one hand, and on the other hand the particular realization, unique and limited to a domain S, known in its totality (after the fact) or more often (in praxi) only partially. This realization is simply the original REV, reinterpreted within a probabilistic framework. But *even within that framework* the RF and its realization are in no way identical, even if the theoretical model is ergodic and stationary, if only because the realization is restricted to a bounded domain S. Regional magnitudes, now interpreted as realizations of random variables, are not identical with their "theoretical" values in the model, which are the corresponding expectations. And it is only regional magnitudes, as calculated from the realization, that represent physical reality and have an objective meaning. Their theoretical counterparts are nothing but conventional parameters, and are not objective, except if, by chance, the particular parameter happens to coincide almost surely in the model with the corresponding regional magnitude.

It is perhaps regrettable that we have had to emphasize the pragmatic aspects of probabilistic modelling, to the detriment of the theoretical aspects. It is undeniable that the problems addressed to in this work, the methods and the models described, appear to be much nearer to the monoscopic and instrumental pole than to the speculative and panscopic one. But manipulation of a model is only efficient if the model is, one way or another, *grosso modo* adapted to reality or at least to that aspect of reality in which we are interested. The specification of our generic models, or more precisely the regional magnitudes that correspond to them (and that we can always study by themselves, independently of the probabilistic interpretation that we choose to attach to them) *also* give expression to structural properties of the real phenomenon. They themselves have a rich physical content. Thus the behaviour of the variogram in the neighbourhood of the origin gives expression to the greater or lesser regularity (continuity) of the spatial variation. The variogram of the theoretical model is, to be sure, devoid of empirical existence, but this is not the case for its regional version. Of course the regional variogram is not given "as is." It results from a first reduction, or modelling, of the phenomenon, carried out by conferring to it the status of REV, after which one performs a manipulation on the numerical data thus contrived (or extracted) from reality. It is real, but its reality is technical rather than natural. It is the result of the application of an algorithm (or a construction). The driving elements of the algorithm come from our physical intuition of the phenomenon, which suggests to us the type of model (whether probabilistic or not etc...) that we shall use. After this, the theoretical model becomes animated with its own life (that of a regulatory fiction), and suggests to us in return the form of the mathematical instruments (the specification) that will be best capable of giving us a precise numerical representation of the physical characteristics that have retained our attention. The choice of model thus plays a capital *heuristic* role, but, when all is said and done, it is only an intermediary. For in the end we shall have to carry out numerical manipulations on numerical data. Whatever the physical intuition and the theoretical interpretations that have led to the choice of these operations, they can and must be analysed on their own. For our problems always eventually reduce to choosing regional magnitudes and attempting to estimate them on the basis of limited information. From that point of view, there can be no miracles: whatever the wealth of our intuition and the subtlety of our models, it is always the same numerical data that we deep kneading and reshaping ad infinitum. The operations we perform cannot, by themselves, generate any additional information. Just as there exists in praxi a threshold of robustness, there must exist a *threshold of realism,* and it is illusory to attempt to overstep it. On the basis of rather poor and limited information, it may be possible to obtain an approximate solution to some simple problems. But, in all honesty, the solution of highly complex problems has to be given up. The ability to recognize the existence of these *thresholds of robustness and realism* and to pinpoint them correctly in each particular case is the earmark of the good practitioner, and this *physical common sense* is, in the final instance, much more important than mathematical theorizing.

An Example of Control of Type Robustness

Here is an example of how one can control the influence of the choice of type of model on the final result aimed at. The test is about the "extension variance" of a point within an interval or a square, that is, the variance of the error committed when we estimate the mean grade over the interval or the square by a single point sample. This variance can be calculated from a model for the covariance or the variogram, whose type must be chosen and whose parameters must be specified. I constructed my example so as to obtain a variogram that has, in the neighbourhood of the origin, an unusual behaviour, and sent the statement reproduced below to various geostatisticians. The successive steps described correspond to actual practice. In particular it is quite customary to request additional drill-holes with a tighter mesh, in order to pinpoint the behaviour of the variogram in the neighbourhood of the origin, and the mine owners usually grant the geostatistician's request in spite of the additional costs involved. This is because a practitioner, even if he is not a geostatistician, can well understand the importance of fine structure, and the necessity to have available information at the appropriate scale. Here is the text of the statement: *"Test of robustness."*

"This is not a trick, but a test to evaluate the robustness of our methods of variogram fitting. I request each recipient to fit a variogram model, and to calculate the corresponding extension variances, on the basis of the following numerical values:

A – Beyond 6, the values differ but little from 1. The unit of length chosen is equal to the drilling mesh (say u = 180 m to fix ideas). The number N of drill-holes is large (several hundreds), so that the experimental points may be considered satisfactory.

$\gamma_1 = 0.63 \qquad \gamma_4 = 0.94$
$\gamma_2 = 0.80 \qquad \gamma_5 = 0.97$
$\gamma_3 = 0.89 \qquad \gamma_6 = 0.99$

I request each of you to carry out the fitting exactly as you would proceed in actual practice, and to do this independently of other statisticians and without being influenced by them.

Once the model is fitted, I would like you to calculate the extension variance of a central point on the unit interval and on the unit square.

B – At the request of the geostatistician, 20 additional drillings with mesh 1/9 (that is, 20 m) are carried out in a representative zone, giving:

$\gamma_{1/9} = 0.25 \qquad \gamma_{4/9} = 0.46$
$\gamma_{2/9} = 0.34 \qquad \gamma_{5/9} = 0.50$
$\gamma_{3/9} = 0.40$

Would you modify your initial fitting? What are the new extension variances?

C – Later on, the mesh is tightened to 90 m, i.e. 1/2 a unit, giving

$\gamma_{1/2} = 0.49$

What are the extension variances for the mesh of width 1/2?

N.B.: These numerical values have been obtained from a model of unusual type. There is therefore a twofold interest in this test: it will bring out on the one hand the dispersion of individual evaluations, and on the other hand their deviations from those derived from the chosen scheme. (The latter are not, of course, the "real" values, since in a real situation there is no guarantee that the particular model is the "real" model.)

In the 10 answers that I received there are notable differences. In general the first model is of the "nugget effect" type, with, in addition, an exponential or spherical scheme (or even 2 superimposed spherical schemes). After including the additional drill-holes, the model becomes richer, becoming either: nugget + spherical + spherical, or nugget + exponential + spherical or nugget + exponential + exponential (or even nugget + 3 spherical schemes) with in any case very different parameters from one model to the other. Other models were also proposed, one of which differing but little from mine. Graphically, however, the fits are quite comparable and lead to values of the extension variance that show very little dispersion.

These values are given below. For the first model, the one fitted without taking into account the additional drill-holes, they are somewhat dispersed and generally rather too large: for in the absence of information about microstructures, the geostatistician tends to choose, in the interests of caution, the hypothesis of a rather large nugget effect, and therefore deliberately overestimates the estimation variances. Once a further fit has been carried out with the help of the additional drill-holes, the models used, although very different from each other, lead to practically the same numerical values.

Model A		Model B (mesh = 1)		Model C (mesh = 1/2)	
—•—	[•]	—•—	[•]	—•—	[•]
0.377	0.405	0.309	0.364	0.231	0.272
0.377	0.412	0.321	0.381	0.243	0.288
0.330	0.370	0.330	0.370	0.232	0.284
0.372	0.405	0.307	0.370	0.227	0.273
0.409	0.430	0.302	0.376	0.232	0.262
0.370	0.405	0.303	0.367	0.225	0.272
0.396	0.430	0.303	0.380	0.227	0.265
0.353	0.388	0.304	0.364	0.225	0.262
0.347	0.372	0.306	0.368	0.237	0.266
0.369	0.402	0.313	0.346	0.232	0.272

Robustness in Relation to Data

In general, an operation is robust if its result is not much influenced by the errors that we might have made either in the collection of experimental data or in the choice of the generic model and/or its specification. There are therefore three types of problems and three types of robustness to consider: the influence of the so-called aberrant data, that of an error in the specification of the type of model, and finally that of an error in the choice of the model type itself. Correlatively, we have data robustness, specification robustness, and type robustness. We have already dealt at length with the latter two, and it now remains to say a few words about the first, that is to examine data robustness and the problem of "aberrant" data.

Among the given data values, some might have been corrupted by gross errors (transcription errors, decimal point in the wrong place, measurements that are wrong for one reason or another) and should, as far as possible, be eliminated. It is often convenient, when there is a large amount of data, to have available automatic procedures to single out the "suspect" data values that are candidates for eventual elimination. Often very simple procedures, such as histograms and scattergrams, will give satisfactory results. But a "suspect" data value is not necessarily wrong or "aberrant" and must never be automatically eliminated. For it may well reflect the real behaviour of the phenomenon (see the example in the following figure).

Whether this kind of behaviour appears to us to be "abnormal" or "aberrant" is a matter of subjective judgement. Reality is what it is, and it may simply

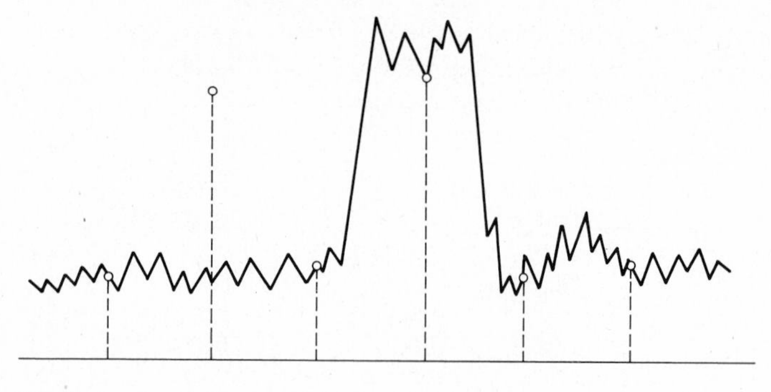

Fig. 2. A suspect data value is not necessarily wrong

be the case that the (more or less stationary) model that we use as a touchstone is unsatisfactory. It is not the data value that is aberrant, but rather the assumption of stationarity that has proved wrong. To eliminate this data value would be a mistake that would subsequently deny us the capability of predicting the occurence of similar accidents in regions for which we have no information.

Data robustness has been the object of very deep study, and the results are no doubt *partly* applicable to the type of models that we have been considering. Besides, we are attempting to protect ourselves not only against aberrant data, but also against the exaggerated influence sometimes exercized on the usual estimators by extreme experimental values (whether very large or very small). This kind of problem tends to manifest itself when our distributions are very far from the Gaussian type. For example, it can be shown that in the case of certain distributions that are *symmetric* around some central value θ, the median is a more robust estimator of θ (i.e. less sensitive to the effect of extreme values) than the arithmetic mean.

However, although such methods can improve data robustness, they may perhaps at the same time weaken model robustness (e.g. we may now have to "assume" that distributions are symmetric etc...). This is a serious objection against the robust, sophisticated estimators that are offered to us by Mathematical Statistics: they often are less robust, relative to the *model*, than simpler estimators. Moreover they are always designed to *estimate a parameter of the model (that is in general not objective)* (e.g. a central value) *and not a regional magnitude*. It is therefore not possible to put them to use as they are in our models. *The general problem of robustness of estimation of a regional magnitude relative to the data, the type and the specification as a whole should be investigated,* but this is not an easy problem.

It is easy to see that an estimator that, like the median, is robust relative to the data is not necessarily adequate to estimate, e.g., the regional mean

$$\bar{z} = (1/S) \int_S z(x)\,dx, .$$

Indeed, since our estimator is of the form

$$z^x = \phi(z_1, \ldots z_N)$$

then if it is optimal in some sense, it must certainly satisfy the following condition: if the mesh determined in S by the experimental points becomes tighter (when their number N tends to infinity, and the experimental points cover in the limit the whole domain), the estimator z^* must converge to the real value $\bar{z}$. Moreover the variable Z^* that is associated with it in the model must converge almost surely to $\bar{Z}$, for when all points are known the best possible estimator of $\bar{z}$ is $\bar{z}$ itself, and it is now experimentally accessible. But clearly an estimator like the median, however robust, does *not* satisfy this condition. It follows that even if the median is really a better estimator that the arithmetic mean when the mesh is wide and the number N of experimental points small, there certainly comes a time when, as the mesh becomes tighter and N larger, the conclusion is reversed: the arithmetic mean gets the upper hand, at least as an estimator of the regional $\bar{z}$.

Notes

[1] In praxi: during action. With a tinge of romanticism we may take it to mean: in the heat of action.

[2] It is not even falsifiable. The concept of stationarity is therefore devoid of objective meaning, not only in the sense of our strong criterion (i.e. decidability), which we shall use for the present, but also in the sense of the weaker Popperian criterion (i.e. falsifiability) to which we shall return later.

[3] or at least falsifiable, when we shall have made our criterion more flexible.

Chapter 5

Sorting out

Once a generic model has been chosen, a *sorting out* operation remains to be carried out. This consists in determining, among the various components and parameters of the model specification, which ones correspond to a reality and which play a purely conventional role. For convenience of exposition I shall distinguish two stages in this quest for objectivity. The first one, which is the *sorting out* operation proper, will be considered in this chapter. It is an internal review of the model that is carried out from the point of view of the model itself. It consists in seeking, among the random variables that the model associates with the various regionals, those which, in the model, coincide almost surely (i.e. with probability one) with their mathematical expectation. These expectations are then objective parameters. This point of view will lead us to consider two important notions. One classical, the notion of *ergodicity*. The other, less classical but no less important for us, will be given the name *microergodicity*. We will of course have to consider as well how things present themselves *in praxi*, that is, attempt to solve the problem of the *estimation* of these objective (regional) magnitudes on the basis of fragmentary information. After this sorting out operation (the internal review of the model) we will proceed to the second stage, when we shall reconstruct the concepts and reformulate the model itself in operational terms. This will be undertaken in Part III.

"Fluctuations" of Regionals

From the point of view of the model, the regionalized variable z defined over the region S is considered as a realization of a random function Z. But this does not in any way mean that epistemological priority has been transferred from the former to the latter. It is still the REV and the regionals that can be derived from it that are the foundation of objectivity. In case of "disagreement" between the model and the REV, it is always the latter that is "right." For the model is only an expression of the decisions (which were perhaps clumsy) and the hypotheses (which were perhaps wrong) that have led us to its choice. For example, in a non-stationary model, the expectation of the RF at the point x

$$m(x) = E[Z(x)]$$

is often interpreted as the "trend" or "drift" of the regionalized variable. This interpretation is sometimes useful. But as long as this concept has not been reconstructed in operational terms it remains a purely conventional characteristic of the model. One should particularly avoid thinking that m(x) represents the "real" value of the REV and that the true numerical value z(x) therefore differs from it only because of an "error of nature". (The situation is, however, somewhat different when the values available in praxi are measurements of an inaccessible REV that are tainted with error.)

What we have said about point values also applies to all regional magnitudes. For consider a given regional $a = \mathfrak{a}(z)$, where $\mathfrak{a}$ is a functional of the regionalized variable z over S. In what follows, we shall mainly consider the two simple examples below:

(1) The (regional) *mean* of the REV over the region S, namely

$$\bar{z} = \frac{1}{S} \int_S z(x)\,dx\,. \tag{1}$$

(2) The *regional variogram* γ_R. It is a function that is equal, for each value of the vector argument h, to half the mean square difference between the values taken by the REV at the two points x and x + h, namely

$$\gamma_R(h) = \frac{1}{2K(h)} \int_{S(h)} [z(x+h) - z(x)]^2 dx\,. \tag{2}$$

Since the two points x and x + h must belong to S, the domain of integration is not S itself, but the intersection of S with the set that is derived from it by the translation −h. We shall denote it by S(h). As for K(h) it represents the measure (the area if the space is two-dimensional, the volume if it is three-dimensional etc...) of the domain of integration S(h). Since it is customary to reserve the use of the symbol γ for the "theoretical" variogram, that is, (in the model "intrinsic RF") for the expectation

$$\gamma(h) = \frac{1}{2} E[(Z(x+h) - Z(x))^2]\,.$$

I have used the subscript R in $\gamma_R(h)$ to underline the fact that we are dealing with the regional magnitude associated with the variogram. One must not say that γ_R constitutes the regional version of γ, because this would be a reversal of epistemological priority. One should rather say that *γ is the theoretical version of γ_R*.

If now in the definition $a = \mathfrak{a}(z)$ we replace the REV z by the RF Z, we obtain a random variable $A = \mathfrak{a}(Z)$ instead of a numerical value a. Within the framework of probabilistic interpretation a is now considered as a realization of the random variable $A = \mathfrak{a}(Z)$ just defined. *This does not mean in any way that the expectation $\alpha = E(A)$ of this R.V. (which is a characteristic of the model) must* *be considered as "truer" or "more real" than the actual value* $a = \mathfrak{a}(z)$ *(the only objective one)*, since this would be the same as saying that reality is wrong and the model is right. In our first example the regional mean $\bar{z}$ is associated with the random variable $\bar{Z}$ whose expectation is equal, if the model is stationary, to a constant $m = E(Z(x))$. Schematically:

$$\bar{z} \rightarrow \bar{Z} = \frac{1}{S} \int_S Z(x)dx\,; \qquad E(\bar{Z}) = m\,.$$

Similarly, associated with the regional variogram γ_R there is, in the model, the random variable Γ_R whose expectation, in the model "intrinsic RF", is equal to the "theoretical" variogram γ:

$$\gamma_R(h) \rightarrow \Gamma_R(h) = \frac{1}{2K(h)} \int_{S(h)} [Z(x+h) - Z(x)]^2 dx\,;$$

$$E[\Gamma_R(h)] = \gamma(h)\,.$$

One must never forget that α, m, $\gamma(h)$ etc... are parameters of the model and depend on the manner in which it has been chosen. They are, indeed, in general *conventional parameters* with no objective meaning. Rigorously speaking, the only case where α is an objective parameter is when, in the model, $A = \alpha$ almost surely, in other words precisely when the theoretical value α coincides with the objective value a. (However, the estimation of α on the basis of fragmentary information is another problem.)

In conformity with habitual usage, let us call the random variable $A - \alpha$ (in the model) the *fluctuation*. The deviation $a - \alpha$ between the regional magnitude and the parameter α in the model can then be interpreted as a realization of the fluctuation. *If the parameter α has an objective meaning, then, by definition, the fluctuation is almost surely equal to zero in the model. Conversely, if the fluctuation is not almost surely equal to zero, this means that the parameter α is not objective.*

Associated with the fluctuation $A - \alpha$, of expectation zero, there are in the model various other characteristics such as its variance $\mathrm{Var}(A - \alpha)$ etc... Let us agree to say that $\mathrm{Var}(A - \alpha)$ is the *fluctuation variance* (in the model) of the regional magnitude. If this variance is large, we must *not* say that "statistical inference" of the parameter α is difficult or awkward but simply that this parameter has very little objective meaning, or none at all. This is because A, or rather its realization a, completely exhausts the objective content that can be attributed to the operation defined by the application of the functional $\mathfrak{a}$ to the REV. In other words, *the fluctuation variance is not a measure of the difficulty of statistical inference of the parameter a of the model, but rather of the lack of objective meaning of that parameter.*

In our first example, if we denote by $\sigma(h)$ the covariance of the stationary model, it can be proved that the fluctuation variance of the local mean is (in the model):

$$\mathrm{Var}(\bar{Z} - m) = \frac{1}{S^2} \int \sigma(h) K(h) dh\,. \tag{3}$$

It is thus only to the extent that this variance is very small that the expectation m (a conventional parameter) takes an (approximate) objective meaning. We shall discuss this point in the paragraph on *ergodicity*.

Similarly, in the intrinsic model, the fluctuation of the regional variogram has a variance whose explicit expression necessarily involves moments of order 4 of the increments of the RF. In the particular case where these increments are Gaussian (i.e. the model "Gaussian intrinsic RF") it can be shown that the fluctuation variance is given by:

$$\operatorname{Var}[\Gamma_R(h)-\gamma(h)] = \frac{1}{2(K(h))^2} \int_{S(h)} \int_{S(h)} \times [\gamma(x-y+h)+\gamma(x-y-h)-2\gamma(x-y)^2]\, dxy\,. \tag{4}$$

If the modulus $|h|$ of the vector h is not small, this fluctuation variance may become extremely large[1]. *This does not mean that statistical inference of the variograms is impossible, but only that the variogram γ has no more than a conventional meaning, apart from a neighbourhood of the origin.* An important consequence of this result is that among the characteristics of the model (e.g. estimation variances) that are a function of the variogram γ, only those that depend in the main on the behaviour of the variogram γ for small $|h|$, and are only weakly affected by its shape outside a small neighbourhood of the origin, may be used to solve real problems (such as, for example, the estimation of z). It is this condition of robustness alone that guarantees the objectivity of the magnitudes we calculate with the help of the variogram. On the other hand, when $|h|$ is small, the variance given by (4) is in general insignificant, and this will enable us to illustrate the concept of *microergodicity*.

Ergodicity

From the classical point of view, the *possibility of "statistical inference"* is always, in the final instance, based on some ergodic property. This is clearly true in the simple case of a trial that is repeated an infinite number of times. If we assume that successive trials are independent, and more importantly that the probability p of success remains constant, then, as we have already seen, the "empirical frequency" will converge to p with probability one. Of the two hypotheses, the first may be weakened: it is enough to assume that the correlation between different trials decreases sufficiently fast with their distance in time. The second one, however, is essential and primarily involves some form of stationarity. It may be rephrased, e.g. by assuming that the probability p(t) of success at time t is itself a stationary and ergodic random function. But in all cases, stationarity will have to be introduced at some stage.

Let us now examine the case of the model "Random Function" and, for example, the "estimation" of the covariance C(x,y). If we are dealing with a repeatable phenomenon, that is, if as many realizations of the random function as we wish are available, then the situation is substantially the same as in the case of the "all or nothing" trials. One can verify (in physical if not in mathematical terms) whether the mean value of the product $Z(x)Z(y)$ converges or not to a limit $C(x,y)$ when the number of realizations increases "indefinitely." It is again not necessary to assume independence of the trials, but only their stationarity in time and their ergodicity. Suppose, for example, that $Z(x, t)$ is the temperature observed at time t at a meteorological station situated at point x in Europe. Let us assume, (this is not obvious, for "secular" variations that are incompatible with a stationary model may intervene), that the mean of the temperatures observed at x every January 1st at midnight tends to a limit when the length of the time series available increases indefinitely. One may also take into account secular variation, as a first approximation, by assuming that its effect is only felt in the long run, and that at the scale of a ten year interval (say 1970–1980), everything occurs as if the temperature were stationary, with a mean $m = m(t)$ that is a very slowly varying function of the time t. This is in essence a "local stationarity" model[2]. We could also have used the classical model "secular (functional) trend + stationary random function". Ergodicity is here an *operational* concept, because the number of observations, although in practice finite, is potentially infinite and increases with time. Thus any deviation from stationarity will certainly be detected in the long run. It may be corrected as a first approximation by fitting a secular trend, but then the residuals will prove to be non-stationary etc.

On the other hand, when we are dealing with a unique phenomenon of bounded extension, such as a mineral deposit, the situation is significantly different since there is no possibility of repetition. In order to evaluate $E(Z_x Z_y)$, assuming that it is possible to give an operational meaning to this expectation, we have at our disposal the sole numerical value $z_x z_y$. We must, therefore, appeal to a stationarity "hypothesis" that will allow us to replace repeatability in time, which is not available, by repetition in space. We shall therefore choose a stationary model, (of course "ergodic" as well) as long as this does not contradict our qualitative information about the phenomenon.

At this point it is necessary to go into a slightly technical digression. Among the various meanings taken by the word "ergodicity" we shall refer only to that according to which spatial means, when taken over increasingly large domains S, converge towards the expectation. But one must clearly distinguish between an ergodic theorem, of general applicability, and an ergodic "hypothesis," which is a constituent part of certain stationary random function models, but not necessarily of all. The ergodic theorem can be stated as follows: Let $Z(x)$ be a stationary random function, and let S_n be a sequence of regions tending to infinity (in the sense that a sphere of arbitrary diameter will be contained in S_n for sufficiently large n). Then the sequence of spatial means

$$Z_n = \frac{1}{S_n} \int_{S_n} Z(x)dx$$

will converge (in a well-defined sense, e.g. "almost surely", or, in the case of RFs of order 2, "in quadratic mean" etc...) to a limit that is independent of the chosen sequence S_n. In particular, the limit is translation invariant. But the theorem does not in any way state that M is identical with the expectation m. In general, this limit, when it exists for "almost all" the possible realizations of the RF, takes different values for different realizations. It is therefore a random variable, not a numerical value. All that we can say is that it is "translation invariant", not that it is constant. On the other hand, in the models for which the ergodic property or "hypothesis" holds, the only translation-invariant random variables are, by definition, the constants and in that case, obviously, $M = m$, or, in other words, the spatial means Z_n converge to their expectation m.

Nonetheless, and whatever its importance in other domains, e.g. in statistical mechanics [3], the distinction between the ergodic theorem and the ergodic "hypothesis" plays no role at all in the case of a unique phenomenon. For we only have, and will never have anything else but *the one realization* of the RF (namely the regionalized variable z) and we shall never know whether the limit M would have taken or not a different value over another realization. The distinction between M and m does not have, in this context, any objective meaning.

It follows from the above that if, for some odd reason, we had chosen for Z a model that does not have the ergodic property, we should hasten to replace it by another model for which this property holds. (It would suffice, for example, to condition the original RF on the random variable M, since this would automatically ensure that $M = m$ without affecting stationarity). This remodelling is equivalent to *choosing* as a definition of the expectation m the limit of the sequence of the spatial means Z_n. If m and M have a purely conventional status, there is no drawback involved in identifying them. If they do have an objective meaning, then it must be the same for both: namely the limit of the spatial means as S tends to infinity.

In any case, to be completely rigorous, this definition can only be conventional. Since the real domain is inescapably limited it is quite impossible to really let S "tend" to infinity, and the notion of expectation m or ergodic limit M, whose definition involves the (fictitious) behaviour of the phenomenon at infinity is, by that very fact, devoid [4] of objective meaning, at least in the strict sense. Nevertheless, one should not be too much of a purist, and if the domain S is in fact quite large, then the variance given by formula (3), although not zero, may be small enough to be considered negligible, and the corresponding parameter in the model, although not rigorously identical with the regional magnitude, may be taken to represent it to a satisfactory degree of approximation. We shall say that in that case ergodicity is practically reached, as far as that parameter is concerned, and this authorizes us to use it legitimately.

Range

From the practical point of view, the fundamental notion here is that of covariance range (also known as "scale of fluctuation"). In approximate terms it is the distance after which correlations vanish or become negligible. *It is only when the domain S is large compared with the covariance range that we can hope to achieve ergodicity in practice* and thus give an objective meaning to the mean m.

An examination of formula (3) will allow us to understand why this is so. In that formula K(h) represents the measure of the domain S(h), which is the intersection of the domain S and its translate by the vector h. In particular K(0) is equal to the measure of S. From this it is easy to see that for large S the variance of $Z-m$ does not differ much from $(1/S)\int \sigma(h)\,dh$. This suggests the following definition: let us call the quantity

$$A=\frac{1}{\sigma^2}\int \sigma(h)\,dh$$

the *integral range*. [We use the traditional symbol σ^2 to denote the variance, in the model, of the stationary RF Z(x). Clearly $\sigma^2=\sigma(0)$, the value at $h=0$ of the covariance $\sigma(h)$.] The quantity A is a length, a surface or a volume depending on whether the space is one, two, or three dimensional. For large S, set $N=S/A$. The quantity N may be thought of as the number of disjoint elements of size A in a tiling of the domain S. In that notation, formula (3) yields the following asymptotic expression for large S:

$$\mathrm{Var}(\bar{Z}-m)=\frac{\sigma^2}{N}. \qquad (5)$$

Thus, as far as the "estimation" of m in this model is concerned, it is as if the estimator Z had been obtained by taking the mean of N independent variables having each a variance σ^2. The integral range is therefore truly the reference element with respect to which it makes sense to say that the domain is large. For the larger the number $N=S/A$, the smaller the variance of $Z-m$ and the more objective the parameter m.

However, a serious objection may be raised here. By defining the integral range through expression (5), that is, through an expression in which the behaviour of the covariance $\sigma(h)$ as $|h|$ tends to infinity intervenes in a crucial manner, we are appealing to characteristics of the model which do not correspond to any objective properties of the real phenomenon. In particular, it is easy to give examples of different covariance functions that are practically identical within the useful domain of h (i.e. when $|h|$ is smaller than the largest diameter of S), but that lead to completely different values of the range A, or perhaps even to an infinite value [for the integral in (5) may diverge for certain choices of the covariance $\sigma(h)$].

To overcome that objection, it is necessary to carry out an operational reconstruction of the concept under attack in the manner previously described.

We introduce at this point a special regional magnitude, which we shall call *the variance of s in S* and denote by v(s/S). We shall not give here its rigorous definition, which is to be found in the specialized literature. But the reader will be easily able to imagine that the given domain S can be divided, at least mentally, into a certain number of elements s of (approximately) equal size. It is then easy to define numerically the variance of the (finite) population of mean values of the REV z over each of the elements s whose union is equal to S. This is, by definition, the variance v(s/S) of s in S, a quantity whose numerical value is a regional magnitude that can be determined experimentally after the fact. Associated with this regional magnitude v(s/S) there is as usual, in the model, a random variable V(s/S) whose expectation $E[V(s/S)] = \sigma^2(s/S)$ may be called, if we wish, the theoretical variance of s in S. The latter quantity is a parameter of the model, which can be easily expressed in terms of the variogram alone. [Explicitly, $\sigma^2(s/S) = \bar{\gamma}(S) - \bar{\gamma}(s)$, where $\bar{\gamma}(s)$, for example, denotes the mean value of $\gamma(x-y)$ when the two points x and y range independently over the domain s.] These expressions may not necessarily have any objective meaning, since the calculation of $\bar{\gamma}(S)$, for example, may involve values of $\gamma(h)$ for h as large as the diameter of S. But, at least after the fact, we can verify whether the equality $v(s/S) = \bar{\gamma}(S) - \bar{\gamma}(s)$ holds approximately or not. Let us note that it is exactly in that way that we have proceeded in Chap. 3 in order to reconstruct the concept of Poisson intensity. It is therefore possible to exhibit, up to a certain approximation, a valid physical law, within some domain of variation of s and S. (Evidently the size of s relative to S must be sufficiently small so that the sample size will be workable: say at least 10.)

The theory now predicts the following: if s (and a fortiori S) is large relative to the range A defined by expression (5), then $\bar{\gamma}(s)$ will differ only very slightly from $\sigma^2 - A/s$. Consequently, if the model is good, we should expect to observe a relationship of the form

$$v(s/S) = A\left(\frac{1}{s} - \frac{1}{S}\right). \tag{6}$$

This relationship is a *physical law* since v is a regional magnitude. Since we can experimentally check (after the fact) that beyond some value of s this law is actually valid, we have, by this very fact, succeeded in constructing the concept of integral range in operational terms. It is now the physical law 6 that constitutes itself the operational definition of the concept.

This reconstruction can be carried out, in principle, only after the fact. But most of the time it is possible, even in praxi, to get an idea of whether the domain S is large enough to assume that ergodicity has been "reached" or "realized." A reliable pointer is given by the shape of the "experimental variogram" (that is, the estimator of the regional variogram that can be calculated, in praxi, from a limited number of sample points). If this experimental variogram reaches a plateau, i.e. a horizontal asymptote that appears stable (apart from residual fluctuations that one always expects to observe), one may, without incurring much risk of error, advance the hypothesis that the range does (objectively) exist

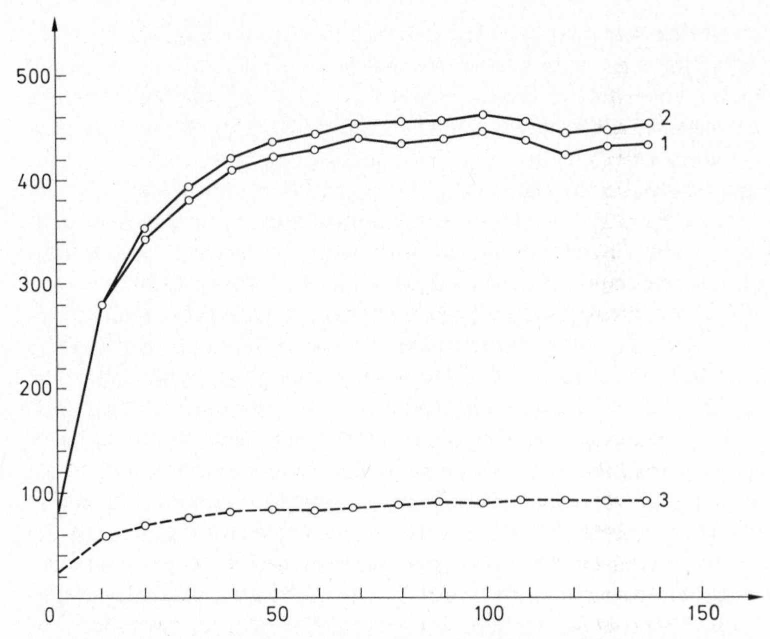

Fig. 1. Variograms associated with features observed on mine facings: 1, chalky concretions; 2, interconcretion minerals; 3, clay joints. [After J. Serra [5]]

and make an approximate measurement of it. Figure 1 is a convincing example of such a situation. On the other hand, we observe in Fig. 2 a peculiar circumstance: the four experimental variograms, constructed in four different directions, coincide more or less up to h=500m. The shape of two of them seems to point to the existence of a plateau and a range of the order of 700m. But beyond h=500m the four curves diverge drastically, and the shape of two of them is completely incompatible with the existence of a plateau. For a practitioner, this is a sure indication of the existence of what is called a "drift", a concept to which we shall return. The fact that the four experimental curves coincide more or less up to the critical value h=500m, after which divergences suddenly appear, is typical of what we shall call local stationarity. We shall give in Chap. 7 the precise definition of the corresponding models, the "local models". But we can already imagine that it is legitimate, in this case, to use locally (i.e. without ever involving simultaneously two points separated by more than 500m) a variogram model whose fit is shown on the right-hand side of the figure.

In Fig. 3 one can see an experimental example of the variance v(s/S) of s in S. Here s (a fixed-size sample) is constant and we have plotted the variances of the grades in larger and larger domains S in a gold deposit as a function of log S/s (from S/s=2 to more than 1000). There is no sign of a horizontal asymptote. Relation (6) is clearly not verified, and there are no grounds to attribute to this phenomenon a finite variance σ^2.

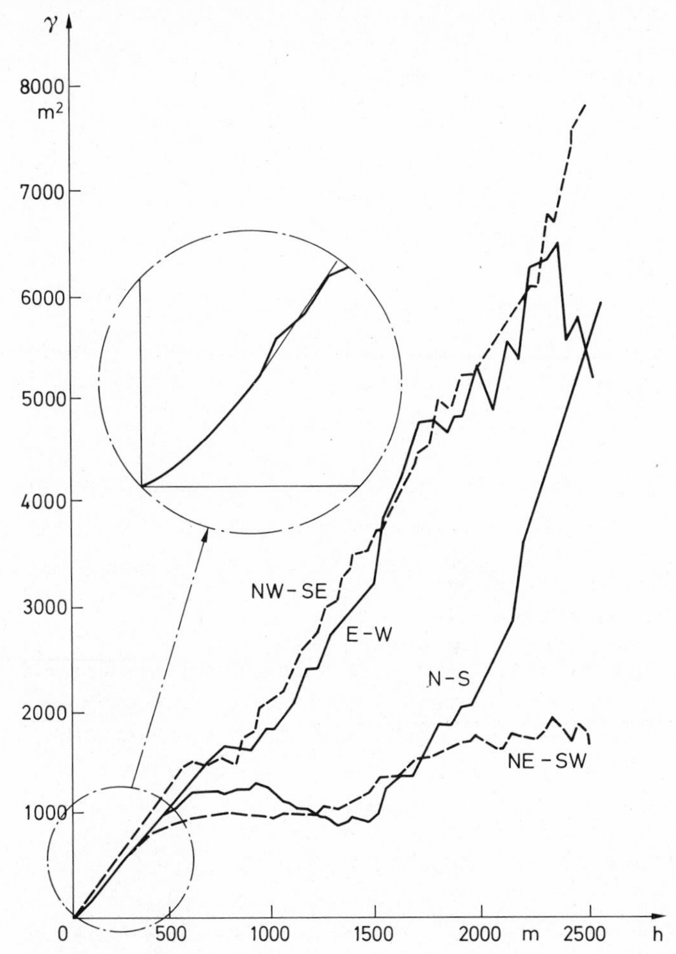

Fig. 2. Experimental variograms of topographic levels (Noiretable fault, 1/25,000). [After J.P. Chiles [6]]

On the other hand, the experimental curve is very well explained by the model:

$$\gamma(h) = \alpha \log|h|$$

(logarithmic, or "Wijsian" variogram). This is a variogram that increases indefinitely and has therefore neither plateau nor range. One may of course surmize that the curve will eventually bend (beyond the furthest observed experimental point) and arbitrarily introduce a plateau (i.e. a finite variance σ^2) and a finite range. But these would be purely conventional parameters, to which no observable reality could be associated. And this would in any case be quite valueless. For as long as the model with a plateau and a range coincided more or less with de Wijs's model up to the largest possible value of $|h|$, it would lead to exactly

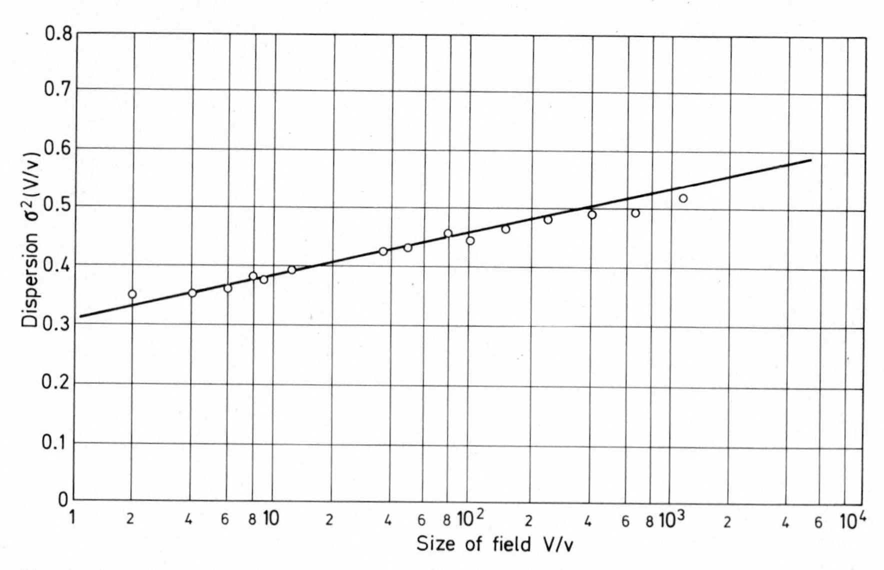

Fig. 3. Experimental variances of gold grade of blocks of fixed size in increasingly large domains in a deposit in South Africa. [After D. G. Krige, quoted in A. Journel [7]]

the same conclusions as the latter, at least as far as the real problems we could raise about the phenomenon were concerned. For in a real problem no distances larger than those that exist between the points of the real object could enter.

Microergodicity

Let us assume that we have measured the values of the regionalized variable z at n points situated, for example, on a square mesh of size a in the domain S (assumed two-dimensional). *There are then two entirely different ways of letting n tend to infinity:*

Either the mesh size a remains constant and we enlarge the domain of investigation, so that the sampling density remains constant, while the size of the sampled area tends to infinity, or else we keep the sampled area fixed, and we reduce the size of the mesh, so that now a tends to zero, and the sampling density tends to infinity. In the first case (assuming certain trivial conditions) we are essentially in the classical "ergodic" situation, which we have already analysed. In the second case, however, we reach in the limit, when $a=0$, the ideal situation that we have called "after the fact", whose epistemological interest we have already noted, and where the numerical value of the REV, interpreted now as a realization of the RF of the model, is known at every point of the domain S. We have dwelt at length on the fact that it is regional magnitudes alone that have an objective meaning, and that therefore any parameter of the model that cannot be identified with one of them is, in the strict sense, meaningless. It will therefore be necessary to perform a sorting out operation among these parameters, and this will be, from the technical point of view, one of the most important aspects of the choice and specification of models. For a given RF model and a given domain S we will have to find out *which parameters of the model are rigorously,* (or approximately, with a given margin of error) *determined by the knowledge of just one realization of the RF over S,* and which are not. Only the former will have an objective meaning and will have an operational definition in terms of "a tends to zero." We shall say that they are *microergodic.* As for the others, since they are devoid of operational meaning, they may to a large extent be chosen arbitrarily, since there will never be any constraining experimental evidence concerning them.

This *microergodic* property is completely distinct from the usual ergodic property, and in my opinion would merit a systematic study. Even in the case of a stationary and ergodic model, neither the mean nor the variance are microergodic. However, *the behaviour of the variogram near $h=0$ is microergodic,* provided, as we shall see, that it is *not* too regular. Moreover, under certain rather weak hypotheses, the microergodic property is preserved even in the non-stationary case. (We are then dealing with the mean variogram, defined as the mean in S of the non-stationary variogram.) Thus stationarity and classical ergodicity do not in any way entail microergodicity, and conversely the latter may hold even in the absence of stationarity.

Indeed, let us examine formula (4) that gives the "fluctuation variance" of the regional variogram $\Gamma_R(h)$ in the model "intrinsic RF". (This formula is only valid in the Gaussian case. In the non-gaussian case, the conclusions may be different.[8]) We have seen that unless $|h|$ is small (and we can now add, unless the phenomenon has a finite range that is small compared with the size of S), this variance can be very large, in which case the behaviour of the theoretical variogram γ for large values of $|h|$ will be devoid of objective meaning.

For small $|h|$, however, there are two possible cases. Suppose, to shorten the argument, that the variogram γ in the model is "isotropic" (that is, that it is a function of the modulus $|h|$ of the vector h alone, and not of its direction), and that for small $|h|$ $\gamma(h)$ is asymptotically equivalent[9] to an expression of the form $a|h|^{\lambda}$. The theory then predicts that only two eventualities are possible:

(a) If $\lambda=2$, then the RF is "differentiable in quadratic mean." This type of model is well adapted to the description of phenomena that are very regular in their spatial variation.

(b) If $\lambda<2$, then the RF is continuous but not differentiable "in quadratic mean", and the model is suitable for the representation of phenomena that are far more irregular and discontinuous. The case $\lambda>2$ is excluded by the theory.

It can then be shown (in the model) that the *variance of the ratio* $\Gamma_R(h)/\gamma(h)$ is, for small $|h|$, asymptotically equivalent to an expression of the form

$$C|h|^{4-2\lambda}+B|h|^n$$

where n is the dimension of the space, namely 1, 2 or 3.

We see that this relative variance tends to zero if and only if λ is less than 2, that is, if the RF is not differentiable in quadratic mean. Furthermore, the following holds:

$$\frac{\Gamma_R(h)}{|h|^\lambda} \to a \tag{7}$$

as $|h|$ tends to zero (in "quadratic mean" and also "almost surely"), provided λ is strictly less than 2, and the parameter a is then microergodic. On the other hand, if $\lambda = 2$, relation (7) does not hold, and microergodicity is no longer assured.

We have thus come to the following very instructive conclusion: the parameters that define the behaviour of the variogram in the neighbourhood of the origin are microergodic, and in particular have an objective meaning, provided only that the phenomenon (or rather the RF that corresponds to it in the model) is not too regular.

This conclusion should not surprise us. Suppose that the domain S is very small (but still contains some open set, e.g. a sphere of arbitrarily small radius).

If $\lambda = 2$, then the realization of Z is a very smooth function, which can be assimilated, over the small domain S, to a parabola for example. The limit a of the ratio $\Gamma_R(h)/h^2$ as h tends to zero depends on the mean value of the slope over the domain S. It may take very different values, depending on the location of S (at S_1, S_2 etc...). In classical terminology, statistical inference is impossible. This really means that the theoretical model has little objective meaning. On the other hand, for $\lambda < 2$, the information contributed by the various points of S (and they are infinite in number, even if they lie very near to each other) is far less redundant, because the RF is far less regular. This leads to microergodicity and therefore to the possibility of determining experimentally the parameters a and λ. Readers who are acquainted with the properties of the Wiener-Levy process will have no difficulties in understanding what has been expressed here only in vague terms.

Let us note, however, that relation (7), for $\lambda < 2$, is an (almost sure) event of the model, but that its counterpart, in terms of regional magnitudes:

$$\frac{\gamma_R(h)}{|h|^\lambda} \to a$$

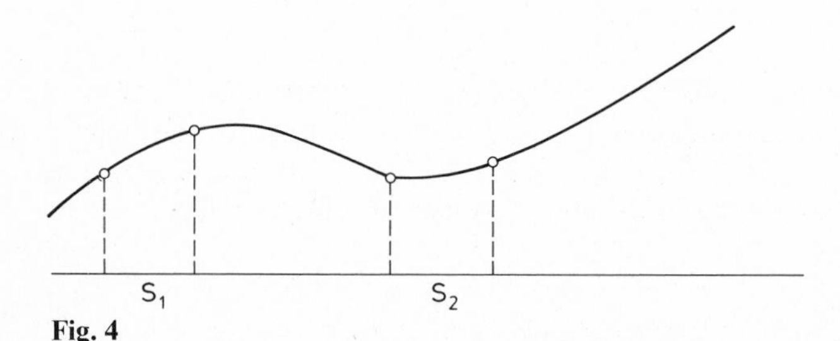

Fig. 4

is not actually experimentally verifiable, since it involves a limit that depends on an infinity of points. Here we are in a case where it is necessary to distinguish between the primary model (the REV) and physical reality.

To pinpoint the exact physical content of the apparently simple concept of "behaviour of the variogram near the origin" it will be necesssary to carry out a full-scale operational reconstruction. We shall return to that point in Part three.

Estimation in Praxi of Regional Magnitudes

The problem of estimation (in praxi) of a regional magnitude on the basis of fragmentary information is a very real problem that constantly confronts the practitioner, and that must not be confused with that of the fluctuation of that regional magnitude around its theorized version (or rather the fluctuation attributed to it by the model). In particular, in the limiting case of perfect information, the estimated value will coincide with the true regional magnitude, while the fluctuation will remain unchanged. In the simplest case, the problem can be stated as follows: knowing the numerical values $z = z(X_\alpha)$ of the REV at the N experimental points x_α, we look for a function $a^* = \mathfrak{a}^*(z_1,\dots,z_N; x_1,\dots,x_N)$ of these variables that will be the best estimator (in a sense to be defined) of the regional magnitude $a = \mathfrak{a}(z)$. In the model, after substituting $Z(x)$ for $z(x)$, a and $\mathfrak{a}^*$ become random variables (not necessarily independent) namely A and A^* respectively. In the *theorized version* of the estimation problem one is therefore led to look for a function $\mathfrak{a}^*$ of N variables such that the random variable $A^* = \mathfrak{a}^*(Z_1,\dots,Z_N)$ is as near as possible (in a sense to be defined) to the random variable to be estimated, namely $A = \mathfrak{a}(Z)$. For example, we might attempt to minimize the estimation variance $\mathrm{Var}(A - A^*)$ under the condition of "zero bias": $E(A - A^*) = 0$. This variance and this expectation are characteristics of the model. From this fact a first limitation follows: *the search for the function* $\mathfrak{a}^*$ *must be restricted to the class* Φ *of (measurable) functions* ϕ *such that the quantities*

$$E[A - \phi(Z_1,\dots,Z_N)] \quad \text{and} \quad \mathrm{Var}[A - \phi(Z_1,\dots,Z_N)]$$

can be expressed in terms of the specification of the model alone. The problem is therefore:

$$\text{Find } \phi \in \Phi \text{ to minimize } \mathrm{Var}[A - \phi(Z_1,\dots,Z_N)]$$
$$\text{subject to the constraint } E[A - \phi(Z_1,\dots,Z_N)] = 0\,.$$

To this obvious limitation (viz. that ϕ should belong to the class of estimators such that we can actually calculate the mean and variance that the model attributes to them) must be added restrictions concerning the objective meaning of this expectation and variance. Since a study of the latter question requires an operational reconstruction of the model, we shall defer it to Part three, and will be content to make here some preliminary remarks. From the practical point of view, the essential point may be summarized in one line:

BEWARE OF OVERSTEPPING THE THRESHOLDS OF REALISM AND ROBUSTNESS!

The best thing to do at this stage is to study some examples.

The Example of Conditional Expectation

Let us consider for the moment the ideal case of a completely specified model, that is, a model where the spatial law of the RF has been explicitly chosen. To each regional magnitude a one can then associate the law of the corresponding random variable A, taken *conditionally* on the available data Z_α, $\alpha=1,2,\ldots,N$. This completely solves the estimation problem, since the best possible estimator is then the conditional expectation $a^*(Z_1,\ldots,Z_N)=E(A|Z_1,\ldots,Z_N)$.

This is of course a purely idealized case, for in practice the available data almost never allow us to specify completely the spatial law of the theoretical RF. So we usually replace the inaccessible conditional expectation by an optimal liner estimator, or by a more elaborate estimator (for example a "disjunctive" estimator, c.f. Chap. 8), that requires fewer prerequisites than conditional expectation.

In reality, however, the reason why conditional laws are not used is not only (in the traditional terminology) because the available data rarely permit the statistical inference of the spatial law. For it does happen sometimes that the model is completely specified, for example in the case of a Gaussian or lognormal RF whose expectation and covariance have already been accurately estimated. Another consideration intervenes, related to what we have called *model robustness* (type robustness and specification robustness). In theoretical (probabilistic) terms, this means asking what would happen if the "real" RF did not completely conform to the chosen model (i.e. if it were not exactly Gaussian or stationary, if the real covariance differed slightly from the one we specified etc...). There is good reason to fear that the conditional expectation will be severely affected by these variations, and that our thresholds of robustness will thus be largely overstepped. We shall return to this point in Chap. 8.

More generally, suppose again that we have succeeded in specifying completely the RF model. Then from the purely theoretical point of view one could think that the best model to use in the estimation problem is that of *the initial RF $Z(x)$ taken conditionally on the available information,* i.e. on $Z(x_\alpha)=z_\alpha$, $\alpha=1,\ldots,N$ fixed. But the most striking feature of the spatial law, once it is thus conditioned, is that, by construction, *it no longer contains any elements that are accessible to direct experimental control:* because now all the experimental data have been included in the conditioning variables, and they appear as parameters in the expression for the conditional law. It is clear that one must really have a blind faith in the chosen model to dare use this conditional law unrestrictedly. There are good grounds to apprehend that not only the *threshold of robustness*, but also the *threshold of realism* will have long since been overstepped.

The Example of Kriging[10]

Let us now look at a less ambitious example, that of the generic model "stationary RF of order 2", whose specification consists of the expectation m and a stationary covariance (around the mean) $\sigma(h)$. Suppose that we wish to estimate the value z(x) of the REV at a point x different from the experimental points $x_\alpha(\alpha=1,2,\ldots,N)$ or, more generally, a "weighted mean" of the type

$$z(p)=\int p(dx)z(x)$$

where p is some measure of unit total weight having its support in S. This is a regional magnitude defined by a *linear* functional of the REV z. The class Φ of possible estimators that can be characterized in terms of the model specification alone consists of *all affine linear combinations* of the form $a+\Sigma\lambda^\alpha z_\alpha$. The theorized version of the estimation problem is thus as follows:

Find constants a and λ^α, $\alpha=1,2,\ldots,N$, that minimize $E[Z(p)-a-\Sigma\lambda^\alpha Z_\alpha]^2$. [We do not need to add the zero bias condition $E(Z^*-Z(p))=0$ because in this formulation it is automatically fulfilled.] The solution is, as is well known,

$$z^x=m+\Sigma\lambda^\alpha(z_\alpha-m)$$

where the coefficients λ^α are the solutions of the system

$$\Sigma\lambda^\alpha\sigma(x_\alpha-x_\beta)=\int p(dx)\sigma(x-x_\beta)\ .$$

This estimator is obviously a function of the chosen specification of m and $\sigma(h)$. Moreover, if the number of experimental points x_α is large, (e.g. N=100 or 1000 etc...) the question of whether the system is "well-conditioned" arises. Finally, if the problem is to estimate z(x) at a given point x, or an average z(p) over the immediate neighbourhood of that point, one can see by examining these formulae that there is a risk of attributing an exaggerated weight to experimental points that lie far away from the point to be estimated. It appears physically implausible that points that are at such a large distance will have such a strong influence. Here again, the blind application of these formulae requires total faith in the model "stationary RF of order 2", Physical common sense must intervene in order to rectify the excessive rigidity of the theoretical model. Of course, reality is approximately "stationary" (in a physical sense that is difficult to make precise), but it is only *approximately* so. We must therefore avoid using estimators that have so little robustness in relation to the stationarity of the theoretical model. In practice, therefore:

(1) we shall impose on the coefficients the additional condition $\Sigma\lambda^\alpha = 1$, and we shall minimize the estimation variance under that constraint. This has the advantage of *filtering out the moment m of order 1* and making it disappear from the expression for the estimator. It will then be unimportant whether m, (whatever its epistemological status), is or is not constant in space. In the disappearance of m one can see again an application of the general rule according to which all operations that are to be actually carried out must be expressed (or expressible) in terms of *regional magnitudes* alone.

(2) More drastically, we shall exclude from our estimator the more distant experimental points, and retain only those that belong to a reasonable neighbourhood of the point to be estimated. This choice of neighbourhood may appear arbitrary, but an experienced practitioner, with a good intuition of the phenomenon, will not hesitate for long. At the cost of a (theoretical) loss of information, which is in any case generally small, the robustness of the estimator in relation to the choice of type of model and specification will be considerably increased. And, incidentally, the amount of calculation will be significantly reduced. (This is not by any means an unimportant consideration in real practice, where the cost of calculations always plays an important role).

Moreover, having thus restricted the class of estimators within which we search for our optimum, we realize that our initial model was *overspecified.* For it has turned out that the conventional parameter m is unnecessary, and that the covariance $\sigma(h)$ can be advantageously replaced by the variogram $\gamma(h)$. Our estimator is now given by the expression

$$z^x = \Sigma\lambda^i z_i$$

where the summation now only ranges over the experimental points x_i that are situated in the neighbourhood of the region to be estimated (e.g. $N = 10$ to 20) and the weights λ^i are the solution of the system

$$\Sigma\lambda^i\gamma(x_i - x_j) = \int p(dx)\gamma(x - x_j) + \mu;\ \Sigma\lambda_i = 1\ .$$

We have, in fact, replaced the initial model "stationary RF of order 2" by the model "intrinsic RF" (IRF) or even, in reality, by the considerably less specified model "locally intrinsic RF" (see Chap. 7). *Thus, we have not jut improved the robustness of our estimator. We have also considerably weakened the anticipatory hypothesis on which it is based.* This is because we have reduced the specification. (The parameter m had disappeared, and it is now necessary to know the variogram $\gamma(h)$ only in the neighbourhood of the region to be estimated, that is, precisely in the range where it has the greatest objective meaning.

To this estimation the model associates an *"estimation variance"* that allows us to gain an idea about the order of magnitude of possible errors. We shall dwell at length in the following chapters on the problem of the objective meaning of this estimation variance and on its reconstruction in operational terms. I shall just give here an example borrowed from Mining Geostatistics [11]. It is about a large copper deposit where 810 panels, all of the same dimension, were estimated by kriging on the basis of neighbouring drill-holes. Since the mesh was regular, the relative geometry of the panels and the drill-holes was invariant, so that the model attributed the same (theoretical) estimation variance to all panels. It was possible in this case after the mining was completed (although this is exceptional in the mining industry) to reconstitute the actual grades of the 810 panels and to calculate the mean error (which was practically zero) and the experimental variance. The results were (in % Cu):

Theoretical variance: 0.117 Experimental variance: 0.118 .

Examples such as this one leave no doubt as to the objective meaning (and the validity) of the concept of estimation variance. But in practice a real difficulty is encountered: the estimation variance, as it can be *evaluated* in praxi, on the basis of partial information and of an approximate estimation of the parameters of the chosen model, is itself already an estimate. Its own variance should be evaluated and so on ad infinitum.

If the model has been correctly chosen, the estimation variance to be evaluated should be expressible, in principle, in terms of the *objective* parameters of the model alone, namely those which are given in terms of regional magnitudes. But the difficulty remains: in order to assess the possibility of estimating the regionals that specify the model, it is in general necessary to bring in working tools other than those that are authorized within the prescribed framework (the specification). We must therefore:

(i) appeal to a *more specified model* than planned,

(ii) *estimate beforehand some regional magnitudes other than those whose estimation we wish to appraise*. For example, in order to assess the estimation of a moment of order 2 (covariance or variogram) and to calculate, say, the corresponding estimation variance, appeal must be made to moments of order 4. Direct estimation of those moments is sometimes possible (but in general conditions are less favourable than for the moments of order 2themselves). Besides, we must then adopt a more specified model (namely one specified by moments of order *2 and 4*) and we now have the problem of assessing this new estimation (that of the moments of order 4). Thus begins a disastrous *infinite regression*. (It is disastrous because experimentally one cannot go very far, and in any case the thresholds of robustness and realism are soon overstepped.) Moreover, direct estimation of moments of order 4 often appears not to be reasonably feasible. One can attempt to sidestep the problem by introducing a much stronger hypothesis that would allow calculation of moments of order 4 from moments of order 2 (e.g. by assuming that the spatial law is Gaussian or lognormal etc...). But this additional hypothesis (which imposes on us the choice of a model that is even more specified than the previous one) must then itself be justified. We are back into a hopeless race towards infinity. But in practice one is not too concerned about this theoretical difficulty, because an error in the estimation variance is far less serious than an error in the estimation itself (and so on).

Notes

[1] I have elaborated upon this point in „Les variables régionalisées et leur estimation", Paris, Masson, 1965, Chap. XIII.
[2] See Chap. 7.
[3] See F. Fer's deep analysis in „L'irréversibilité, fondement de la stabilité du monde physique", Paris, Dunod, 1977.
[4] When we shall carry out in Part III the operational reconstruction of our models, the situation will be very different. The parameter m will then be *defined* as a certain spatial average, and its objectivity will thus be automatically guaranteed.
[5] „Morphologie Mathématique et genèse des concrétions carbonatées des minerais de fer de Lorraine", Sedimentology, 10, (1968), 183–208.
[6] „Géostatistique des phénomènes non stationnaires dans le plan". Thesis, Nancy, 1977.
[7] Journel A. B. and Huijbregts C. J. "Mining Geostatistics," New York, Academic Press (1978).
[8] For the non-gaussian case, in Alfaro-Sironvalle, M. "Etude de la robustesse des simulations de fonctions aleatòires" – Thesis, Fontaineblau 1979.
[9] In other words, $\gamma(h)/|h|^{\lambda}$ tends to a when $|h|$ tends to zero.
[10] This term, derived from Mining Geostatistics, alludes to the work of D. G. Krige, and refers to the "best linear estimator verifying such and such a condition of unbiasedness".
[11] After A. Maréchal and I. Ugarte, quoted in Journel op.cit.

Part III Operational Reconstruction

Chapter 6

Global Models

In the last three chapters of this work, I shall attempt to throw some light on the physical content that is implicit in the probabilistic models we use to represent regionalized phenomena. The richness of a physical concept is measured by the number and variety of the phenomena between which it highlights similarities and therefore by the richness of the network of relationships and physical laws in which it is involved. It is this network that, strictly speaking, constitutes the operational definition of the concept. We shall therefore have to carry out a full-scale *operational reconstruction* of our probabilistic models. It is an immense task, and there is no question of being able to carry it out in its entirety. I shall content myself with examining some typical particular cases, without even claiming in any way that the procedure I shall follow (namely going through the intermediary of "probabilistic representations") is the only possible one, or even necessarily the best.

We shall dwell at length on the example of the variogram. Let us make it clear at the outset that we are not talking about the variogram of the model, γ, whose objective status is still partly in doubt, but about what we have called the regional variogram γ_R – not the expectation, but the mean value in physical space of the square of the difference $z(x+h)-z(x)$. Let us write again its definition, that is, formula (2) of the previous chapter:

$$\gamma_R(h)=\frac{1}{2K(h)}\int_{S(h)}[z(x+h)-z(x)]^2dx\,. \tag{1}$$

In this formulation, there is no suggestion of probability. The regional magnitude γ_R is a physical reality that we can examine by itself, independently of the interpretations (whether probabilistic or not) that we shall associate with it later. It is a sort of summary of the structural characteristics of the REV and conveys a wealth of information of physical, not probabilistic, nature to us. We have already seen how much its behaviour for large values of $|h|$ reveals to a practitioner. If the curves observed along different directions of space stabilize around a common horizontal asymptote (or plateau), we have a sure indication of "stationarity" (in the physical sense), that is, of the homogeneity of the phenomenon in space. To this physical reality is associated an operational concept, that of range, whose construction from a physical law (the variance of s in S) we have already performed. If, on the other hand, the curves diverge violently in the various directions, then we say that there is a "drift." This is a more difficult concept to delimit, and we shall return to it later. But it does correspond to some form of physical reality, to a certain type of behaviour of the spatial variation of the REV.

The behaviour of the variogram near the origin proves to be perhaps even more revealing. By its very definition, γ_R characterizes the regularity and continuity properties of the phenomenon. Or, more precisely, it provides us with an average picture of these properties. The faster its growth, as $|h|$ increases, and the faster will the "influence" of a point x on its neighbour $x+h$ decrease, on the average, and therefore the more discontinuous and irregular will the phenomenon itself prove to be. We are led to represent the behaviour of γ_R around $h=0$ by simple models, which are really just approximations, to be used only at a certain scale of operations: for example parabolic behaviour ($\alpha|h|^2$), or linear behaviour ($\alpha|h|$), or "nugget effect" (a discontinuity at the origin). These different types of behaviour correspond to real properties of the phenomenon. In the parabolic case, we are dealing with a very regular phenomenon, which we can *map*, at that scale, in all its details. In the linear case, mapping is still roughly possible, but overlooks many details on account of the overlarge local irregularity. Finally, if there is a nugget effect, we know in advance that any map will be largely illusory. These purely empirical observations force themselves on the practitioner with an inescapable obviousness and necessity. There are at work physical laws that although not yet formalized, will allow us to carry out the operational reconstruction of the concept of "shape of the variogram around the origin."

In order to proceed further with the analysis, we shall introduce an intermediate step. Between the initial REV and the probabilistic model we shall interpose what we shall call "probabilistic representations". These will be completely objective models whose only purpose is to present in probabilistic form the totality of the information contained in the REV. They are tautological, accessible only after the fact, and therefore useless as they stand. Their usefulness is purely methodological. They provide criteria for assessing the simpler models used in practice (stationary, intrinsic etc...). The latter then no longer just fall off from a Garden of Eden of ideas. Their status is now that of *anticipated simple approximations* of the inaccessible, strictly objective models that we have called "representations." Moreover, the mysterious problem of "statistical inference" will also appear in a somewhat different light. For as soon as the essential part of the physical reality that we are attempting to account for is synthesized in the regional variogram, the problem is no longer to "infer" the mysterious γ of the theoretical model. It becomes simply the problem of estimating the spatial integral that appears in formula (1), on the basis of the data available to us in praxi. *We are no longer carrying out a "statistical inference", but an approximate calculation of an integral* on the basis of a small number of experimental points. This, of course, does not solve the problem, but the obscurity that concealed its real nature has now been dispelled to some extent.

In the next paragraph we shall define two kinds of probabilistic representations of a REV: *transitive representations,* which are useful for studying global problems (e.g. the estimation of the integral $Q=\int z(x)dx$, and *sliding representations,* which will enable us to tackle local problems [e.g. the estimation of the value z(x)at a given point]. The remainder of this chapter will be devoted to global problems and models, while local models will be treated in the next chapter.

Representations: Strictly Objective Probabilistic Models

The extremely strict objectivity principle that we have adopted raises the question of whether it is at all possible, for a given REV, to set up a probabilistic model that is perfectly objective after the fact, i.e. such that all its parameters can be identified with regional magnitudes. Besides, this principle is perhaps an excessive requirement, for we are only seeking monoscopic models, i.e. models that have been set up specifically in order to solve a particular problem, by means of a method that we have chosen in advance. The definition of the problem and the method chosen usually involve certain parameters that are independent of the REV, but are unequivocally determined. We shall call them *methodological parameters*. For example, suppose that the problem is to estimate the integral $Q=\int z(x)dx$ of a REV z(x) in two-dimensional space, on the basis of data $z(x_\alpha)$ obtained by sampling a regular grid of experimental points x_α located on a square mesh $(a \times a)$. The parameter a (the mesh size) is not related to the REV, but is imposed on us by the problem to be solved. It is a methodological parameter, which can be incorporated into the monoscopic model as an auxiliary parameter without any arbitrariness.

Let us agree to say that a model is *strictly objective* if its specification involves only *objective parameters* (i.e. parameters that can be identified with regional magnitudes) and *methodological parameters* (that have been imposed unambiguously by the problem and the method chosen to solve it), to the exclusion of any kind of conventional parameter. We shall see that it is indeed always possible, *after the fact,* to construct strictly objective models, which *would* have enabled us to give a perfect solution to the problem, had it been possible to specify them correctly *in praxi*. There are of course no miracles, and the perfect specification of an objective model remains out of reach, in praxi, precisely because we do not know the exact values of the regionals associated with the objective parameters. Theoretically, this is a vicious circle. And it is at any rate clear that a strictly objective model – precisely because it is strictly objective – can only be *tautological*. It just presents the same information, that which is contained in the given REV, and which is accessible only after the fact, but in a different form. It is simply a *representation* of the original REV.

In practice, however, the introduction of certain reasonable approximating and simplifying hypotheses allows us to break out of the circle. We meet again here with our anticipatory hypotheses (with all the fertility and vulnerability that they imply) and we can now attempt to make more precise their epistemological status, which is that of an *anticipated approximation*. To the rigorous representation of the REV (a model too rich to specify in praxi) we shall substitute an approximate, much simpler, one, namely a type of model whose parameters can be reasonably estimated in praxi, but that nevertheless leads to a solution of the problem that is equivalent in practice to the one we would have chosen, had the rigorous representation been available. Of course, as always, it is only after the fact that we can assert this practical equivalence. The risk of error has not been eliminated: it has just been circumscribed.

Let us note that the "representation" (or strictly objective model) has a probabilistic nature, while the real situation is unique and has nothing random about it, at least at the outset. The representation is chracterized by two elements: on the one hand the REV z, defined over a domain S, which is known only after the fact, and on the other hand the grid of experimental points x_α, $\alpha=1,2,\dots,N$ at which the numerical values $z_\alpha=z(x_\alpha)$ are given in praxi. In order to carry out probabilization, we must somehow surround the real situation with a halo of virtual states that will be considered as "possible" and assert in some way that the present state could be any one of those possible states, and does not stand out on account of any peculiarity. The model will then be completed by specifying a probability law over the set of states that are considered possible. Here the main danger to avoid is clearly arbitrariness. If we want to remain completely objective, then, in the definition and probabilization of these virtual states, we may only utilize the REV itself and the structure of the information grid. But we must at the same time give back a certain relative mobility to these two elements in order to generate a family of realizable virtual states. The basic idea, which is very simple, is to imagine that the information grid is displaced (or deformed) in the domain S of the REV.

To fix ideas, let us consider the simplest possible case, namely one where the experimental points x_α form a regular grid, with a square of rectangular mesh, in two-dimensional space, and where the problem is that of *linear estimation*. We must consider the cases of global and local estimation separately.

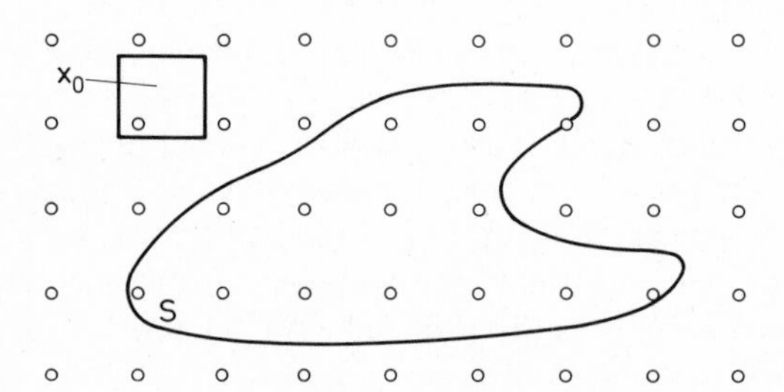

Fig. 1. Global estimation

(a) *Global estimation:* The problem is, for example, to estimate the integral

$$Q = \int z(x)dx .$$

The exact shape of the domain S of the REV is unknown (in praxi), but the problem of global estimation can be reasonably attempted, in practice, only if the regular grid of real points x_α, of finite number, actually extends beyond the domain S so that it can be continued to infinity (by setting the values of the additional fictitious points to zero) without any drawback. The following method will then yield a *transitive representation*[1]: one of the experimental points, say x_0, is chosen arbitrarily to be the origin of the grid, and is displaced within the rectangle defined by its mesh, carrying along the whole mesh. Its displacement defines the set of virtual states, and it only remains to consider the point x_0 as a random point uniformly distributed in the mesh rectangle to complete the definition of the model.

(b) In a *local estimation* problem, on the other hand, the problem is to estimate a shape that is *tied to the sampling grid,* for example, the mean value of z(x) in the "zone of influence" $\Pi(x_0)$ of a point x_0 of the grid (that is, the mesh rectangle centered at the point x_0). In view of the robustness considerations previously discussed, it has been decided to base the estimation only on the nearest experimental points, e.g. the point x_0 itself and the its eight immediate neighbours $x_1,\ldots,x_8$.

Under these conditions, we shall let the complete configuration formed by $\Pi(x_0)$ together with the nine points $x_0,x_1,\ldots,x_8$ move in space. The point x_0, carrying the whole configuration, will range over some domain S_0 (chosen by us: S_0 is in general a subdomain of the total domain S, chosen in praxi in the light of the sampling itself). It then suffices to consider x_0 as a random point uniformly distributed in S_0 to complete probabilization. This second method corresponds to what we shall call *sliding representations*. In the first method, the zone to be estimated (which is identical with the whole domain S) remains fixed in space, while the sampling grid is displaced by translation. (In mining terminology, this is called "fixed domain and floating implantation"). In the second one it is the complete configuration, the experimental points as well as the zone to be estimated, that is displaced as a whole (in mining terminology one speaks

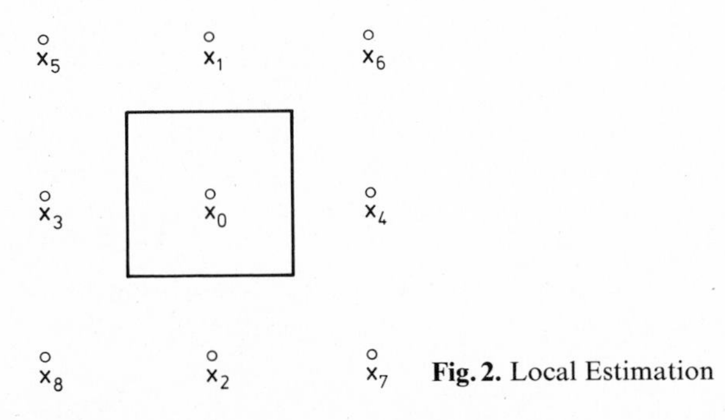

Fig. 2. Local Estimation

of a "floating domain and preferential implantation", it being understood that the floating domain corresponds to the zone to be estimated $\Pi(x_0)$ and not to the whole domain S that evidently remains fixed in space).

Transitive Representations

Apart from the REV itself, defined over an (unknown) domain S and continued to infinity by setting $z(x)=0$ when x does not belong to S, we consider a grid of "sampling points" $x_i=x_0+h_i$ situated on a regular mesh. To fix ideas, let us take the case of a rectangular mesh in two-dimensional space. (We can, of course, consider other types of mesh, and also treat the three-dimensional case). The mesh is defined by a base rectangle Π, of sides a_1 and a_2, centerd at the origin of coördinates. The components along the two coordinate axes of the vectors h_i that define the mesh are integral multiples of a_1 and a_2 respectively. The origin x_0 of the grid is "positioned at random in Π". (This means that we take as our sample space the rectangle Π, and as probability P the uniform law over Π.) The experimental data $z(x_\alpha)$, positioned at the points $x_i=x_0+h_i$ become themselves, in the model, random variables (for they are functions of $\omega=x_0$).

This scheme is a good representation of the situation (frequently encountered in practice) when we do not know (in praxi) the exact boundaries of the domain S of the REV, and where the whole grid of experimental points has been positioned "just somewhere" relatively to that field. The space of virtual states is made up of all the possible positionings of the one grid, taken as equally probable. (This leads to the uniform density over Π.) This probabilization represents, in essence, a random translation of the whole grid. One may think that it would be possible to "enrich" the model by adding a random rotation. But this would not be realistic, because if there exist preferential directions in the spatial variation of the phenomenon, they will, in general, be easily detected experimentally as soon as we have the data from one regular grid. It would therefore be quite incorrect to consider, among all virtual states, all possible orientations of the grid as equally probable (since one has in fact already been realized and we know it). As far as translations of the origin of the grid is concerned, however, the situation is basically different. One may of course think that certain positionings of the point x_0 will lead to exceptionally favourable or unfavourable results, but it impossible (in praxi) to locate them. It is therefore "natural" to consider all the possible positionings as equally probable. This is, of course, a methodological choice but it is unequivocally dictated to us by the very nature of the problem (global estimation) and the structure of the information (a regular grid positioned without precise information about the boundaries of the domain).

The magnitude to be estimated $[Q=\int z(x)dx]$ is a regional magnitude possessing a global character. The simplest linear estimator that can be imagined is

$$Q^*(x_0)=v\Sigma z(x_0+h_i)\ , \qquad (2)$$

where v is the measure of the basic element Π (length, area or volume, depending on whether the space is 1, 2 or 3-dimensional). Since x_0 is (in the model) chosen at random in Π, the estimator $Q^*(x_0)$ is a random variable defined over the space $\Omega = \Pi$ equipped with the uniform probability over Π. It follows from the way that this random variable has been defined that all its properties must be expressible exclusively in terms of the characterization of Π (methodological parameters) and the REV z (objective parameters). Thus we have actually set up a strictly objective model (a "representation").

The expectation of $Q^*(x_0)$ is, by definition,

$$E[Q^*] = \frac{1}{v} \int_\pi Q^*(x_0) dx_0 .$$

If we replace $Q^*(x_0)$ by its expression 2, it is easy to see that summation over all the vectors h_i of the mesh, together with integration over the base rectangle, will reconstitute the whole integral, extended over the whole space, so that we are left with

$$E(Q^*) = \int z(x) dx = Q .$$

Thus we have, in the model, an *unbiased* estimator of the (objective) parameter Q that we are trying to estimate. A similar calculations shows that the moment of order 2, that is, the expectation of $(Q^*)^2$ is

$$E[(Q^*)^2] = v \sum_i \int z(x) z(x+h_i) dx .$$

We are thus led to associate to our REV z a function g, called the *transitive covariogram*, defined by the relation

$$g(h) = \int z(x) z(x+h) dx . \tag{3}$$

The moment of order 2 can then simply be written

$$E[(Q^*)^2] = v \Sigma g(h_i) .$$

If we integrate with respect to h the function defined by (3), we obtain

$$\int g(h) dh = Q^2 .$$

The right-hand side of this expression is the square of the quantity $Q = \int z(x) dx$ that we are trying to estimate. Let us recall, however, that in the model the random variable Q^* has a variance $\mathrm{Var} Q^* = E[(Q^*)^2] - Q^2$ that we call the *estimation variance*. From the preceding results, we see that it is given by the expression

$$\mathrm{Var} Q^* = v \sum_i g(h_i) - \int g(h) dh . \tag{4}$$

This formula deserves some comments. It connects an objective, unequivocally defined, concept namely the *estimation variance*, with a very simple concept, namely the *transitive covariogram* g. The latter, as can be seen from formula (3), is so to speak the "deterministic" equivalent of the covariance $C(h) = E[Z(x)Z(x+h)]$ of a stationary random function Z. At any rate, it will perform the same role in practice. But it has not been necessary to introduce any "hypothesis" of stationarity concerning the REV. In fact, the latter has not at any stage been interpreted as the realization of a random function. The stationarity property that appears here is not a property of the phenomenon, but exclusively a characteristic of our sampling grid. Our transitive covariogram contains exactly the same structural information as a stationary covariance, but possesses a decisive advantage over the latter: it is unambiguously defined by relation (3), and its meaning is purely objective. If we are content, as is often done in practice, to characterize the precision of the global estimation by the estimation variance alone, then formula (4) shows that we have a well-formulated problem that has an unequivocally defined solution, at least after the fact.

Besides, the physical meaning of formula (4) is simple and easy to grasp. Indeed, the estimation variance is expressed as the difference between the approximate and the exact value of the same integral $\int g(h)dh$. It can therefore be thought of as the error committed by approximating the integral with a discrete summation over the points h_i of the mesh. In view of what is known about the numerical evaluation of integrals, we must therefore expect that the variance will decrease as the mesh becomes tighter, (this is pretty obvious), and also as the function g becomes more regular. Since the regularity of g reflects the average continuity properties of the regional variable z itself, we conclude that for a given mesh size the estimation will be the more precise the more regular the REV itself in its spatial variation. This conforms well with what intuition suggests, and also expresses this intuition in a rigorous mathematical form.

Estimation of the Transitive Covariogram

It remains to examine how matters present themselves in praxi. Since the methodological parameters (v and the vectors h_i) are given at the outset, the only outstanding problem (albeit a serious one) is that of the *estimation of the transitive covariogram* g.

Indeed, at first glance, this estimation appears to raise problems that are more difficult than the initial problem (namely the estimation of Q). It comprises two parts: on the one hand, the estimation of the values $g(h_i)$ of the covariogram for the arguments h_i corresponding to the vectors of the grid (the only values for which we have experimental information that is directly usable); and on the other hand an interpolation between the estimated values, to obtain an evaluation of g(h) for the remaining vectors h.

For a given vector h_i the estimation of $g(h_i)=\int z(x)z(x+h_i)dx$ is analogous to that of Q, since the REV z(x) is simply replaced by $z(x)z(x+h_i)$, and one is therefore led to take as its estimator

$$g^*(h_i)=v\sum_j z(x_0+h_j)z(x_0+h_i+h_j)\,.$$

Indeed, on account of the translation invariance of the grid, the points $(x_0+h_i+h_j)$ are experimentally available, just as the (x_0+h_j) are. For fixed h_i, the product $(z(x)z(x+h_i)$ is itself a function of x, that is, a REV. However, the domain of this new REV is the intersection $S \cap S\text{-}h_i$, which is smaller than the initial domain, and the number of useful data points (i.e. those different from zero) is therefore smaller than the number available for estimating Q. One might think that the precision of the estimation could be assessed by applying formula (4) to the covariogram associated with the product REV $z(x)z(x+h_i)$, and so on. But this would lead us into a disastrous infinite regression, which would require covariograms of higher and higher order, to be estimated from less and less data (because of the shrinking of the useful domain), and in any case under more and more doubtful conditions (for it is well known to statisticians that the estimation of moments of even moderately high order soon loses all robustness). It is therefore not possible to go very far in that direction and, in practice, one will often give up any attempt to associate a variance with the estimator $g^*(h_i)$.

The second problem is that of the interpolation of g between the (estimated) values $g^*(h_i)$. One might think that since, according to formula (4), we apparently need only the integral $\int g(h)dh = Q^2$ all we need to do is replace it with its estimator $(Q^*)^2$.

Unfortunately this would attribute to Var Q^* the value zero, as follows from the easily verified identity

$$v\Sigma g^*(h_i) - (Q^*)^2 = 0 \ .$$

This result is easily understood. When we replace Q and the g(h) by their estimators Q^* and $g^*(h_i)$ obtained from the grid of points $x_0 + h_i$, we replace the problem of estimating the integral $Q = \int z(x)dx$ by that of estimating the discrete sum $Q^* = v\Sigma z(x_0 + h_i)$, and the above identity simply tells us that $Q^*(x_0)$ is an excellent estimator of $Q^*(x_0)$ itself. In less trivial terms, this means that it is not possible to derive from the same set of data both an estimator and the precision of that estimator.

The conclusion is (as always) that there is no option but to introduce an approximating anticipatory hypothesis: we replace the real (unknown) g(h) by a function $\bar{g}(h; \lambda, \mu)$ of a suitable chosen type, which depends on a small number of parameters λ, μ, ..., to be specified as best as we can from the experimental data $g^*(h_i)$. This choice is of crucial importance, since, in the end, the result we obtain by applying formula (4) to the model $\bar{g}$ (once specified) will be of value only insofar as the simplified model represents an acceptable approximation of the real unknown g(h). The worst thing one could do here is to rely blindly on some automatic process to interpolate between the experimental values $g^*(h_i)$.

To make an informed choice, it is necessary to look more closely at the structure of formula (4), in order to identify, as far as possible, the factors that have the greatest influence on the estimation variance. According to its definition (3), the transitive covariogram g is a "positive definite" function. Indeed, this fact guarantees that expression (4) will always be positive, as a variance should be. The model $\bar{g}$ to be chosen will therefore have to be a function of that type. Now the fact is that the irregularities of this type of function are mainly confined to a *neighbourhood of the origin*. Our attention is thus attracted again to the behaviour of the covariogram around $h=0$. Moreover, a detailed analysis shows that the numerical value given by formula (4) depends principally on the behaviour of g(h) in a neighbourhood of the origin whose dimensions are comparable to those of the base element Π of the sampling grid. The crucial point is therefore the interpolation of $\bar{g}(h)$ between $g^*(0)$ and the first experimental point available in the neighbourhood of zero, and in particular the type of analytic behaviour exhibited by the chosen model $\bar{g}$ in that region (for example: linear with nugget effect or $\propto |h|^\lambda$, $0<\lambda\leqq 2$). It is the choice of the type of behaviour of $\bar{g}(h)$ in the neighbourhood of $h=0$ that constitutes the essential part of the anticipatory hypothesis.

The Approximation Formulae

It is of great interest, from the physical point of view, to study the behaviour of the estimation variance when the mesh size becomes very small, and to relate it to that of the covariogram g around $h=0$. Let us consider first the case of one-dimensional space (the straight line) and let L (instead of S) denote the extent of the domain of the REV. The length L is also the *range* of g(h), that is, the distance beyond which that function becomes identically zero. For $h=L$, g(h) exhibits some analytical irregularities which depend on the more or less brutal manner in which g(h) joins the h-axis there. Let a denote the mesh size, that is, the constant distance between the successive sampling points, and $\sigma^2(a)$ the estimation variance given by formula (4). It can then be shown that this variance is the sum of two terms:

$$\sigma^2(a)=T(a)+T'(a)\,.$$

The first term, T(a), is tied to the behaviour of g around $h=0$; the second, T′(a), to its behaviour around the range. The function T′(a) essentially depends on the quantity $\varepsilon=L/a$ modulo 1 [that is, $\varepsilon=(L-na)/a$, where n is an integer such that $na\leqq L<(n+1)a$]. It can be shown that its mean value, when ε varies from 0 to 1, is equal to zero. For that reason it is called the *fluctuating term* (or Zitterbewegung). Its amplitude may be considerable, as shown in Fig. 3.

Altough it is possible to write down a theoretical expression for the fluctuating term, it is never possible, in applications, to predict its exact value since the exact value of the range L is only known with a precision of $\pm a$ and therefore ε, which is equal to L/a modulo 1, is completely indeterminate. We therefore rely on the fact that the mean value of T′(a), as a function of ε, is zero and simply neglect it. Let us note the disguised probabilistic character of this approximation. It is as if the variance $\sigma^2(a)$ were a random variable made up of two parts: an expectation equal to the regular term T(a), and a purely random and unpredictable part represented by the fluctuating term T′(a). In fact, Fig. 3 calls to mind the pre-random situation analysed by J. Ullmo, where inseparable initial

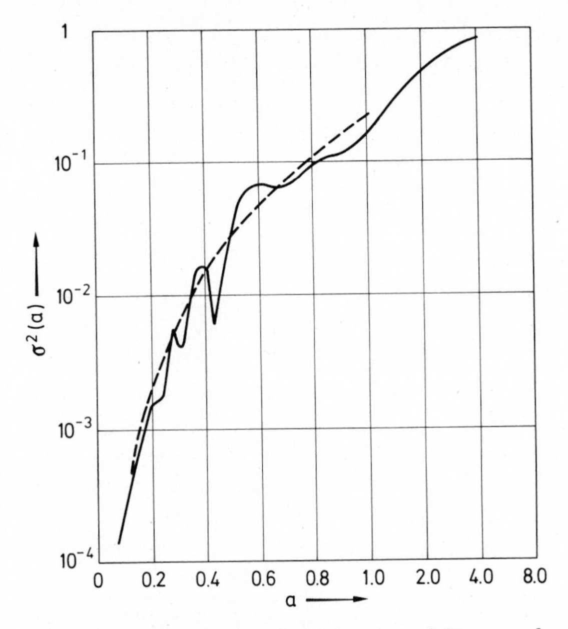

Fig. 3. Zitterbewegung in the estimation of the area of a circle of unit diameter by means of a grid with square mesh. The abscissa is the mesh size a. The ordinate is the corresponding estimation variance $\sigma^2(a)$. The full line represents the exact value $T(a)+T'(a)$. The dotted line represents the regular term only

conditions lead later on to a separation of the observed phenomena (see Chap. 1).

As far as the *regular term $T(a)$* is concerned, it is tied, as already stated, to the behaviour of g(h) around h=0. For example, in a model where g(h) is of the form $g(h)=g(0)-A|h|^{\lambda}+\ldots(0<\lambda\leqq 2)$, we find, for small mesh size a, an asymptotic expression of the form

$$T(a)=AT_{\lambda}a^{1+\lambda}$$

(where T_{λ} is a known numerical coefficient. For example, $T_{\lambda}=1/6$ if $\lambda=1$ etc...). Let n denote the number of sampling points x_i where $z(x_i)$ is not zero. We have approximately $n=L/a$. Replacing a by L/n, we see that for large n the above formula leads us to predict an estimation variance of the form

$$\sigma^2(a)=\frac{C}{n^{1+\lambda}}\,. \tag{5}$$

In the case of a variogram that is "linear" near the origin (a frequently occurring case), i.e. for $\lambda=1$, it follows that *the variance varies inversely as the square of the number of useful samples,* at least when n is large.

In the case of *two-dimensional* space and a rectangular mesh $a_1 x a_2$, we obtain similar results. There is still a fluctuating term that is not negligible, as shown in Fig. 3, and a regular term, now of degree $2+\lambda$. [More precisely, if $a_1 \leqq a_2$, it contains a term in $(a_2)^{2+\lambda}$ and another in $a_2(a_1)^{1+\lambda}$.] If we take, for example, the case of a *square mesh* $a_1 = a_2 = a$ then the variance is proportional to a. Since the number of useful data points is approximately $n = S/a$, we obtain a law of the form

$$\sigma^2(a) = \frac{C}{n^{1+\frac{\lambda}{2}}}. \qquad (6)$$

For example, in the frequently occurring case when λ equals 1, *the estimation variance now varies inversely as the power 3/2 of the number of useful data points.*

Equations (5) and (6) are in fact *physical laws*. Indeed, it is possible to verify them experimentally, at least after the fact. They show to us how the *order of magnitude* of the estimation variance depends on the number of samples. And since they are physical laws, the parameter λ that appears in them assumes an operational status. These laws assure us that the behavioural model that we have attributed to the covariogram is more than just a graphical approximation. It is a physical 'concept (the behavour of g near the origin, represented by a *"regularity exponent"* λ) since the physical laws that make up that concept allow us to measure it.

The Case of an Irregular Mesh

The transitive covariogram g(h) that appears naturally in formula (4) for the estimation variance is formally very analogous to a stationary covariance, and the problem of its estimation in praxi is strongly reminiscent of the difficulties raised by the "statistical inference" of a covariance function from a unique realization of a stationary random function. Here, however, we have not introduced any hypothesis about the stationarity of the REV, which has not even been considered as a realization of a random function, and has retained its "deterministic" status. On the other hand, the regular sampling grid, with its origin x_0 uniformly distributed over the base element Π, has been treated as a stationary point process. It is the *stationarity of the grid* that has allowed us, in the absence of any hypothesis about the REV, to benefit from the highly advantageous conditions that are usually thought to be due to the stationarity of the phenomenon itself, or of the REV that represents it. We shall return later to this important remark: *stationarity may often be introduced, not as a hypothesis bearing on physical reality, but as a characteristic of the method of estimation that we have chosen.*

Let us now consider the case when the experimental points are positioned in an arbitrary way in the domain S and its vicinity, and do not form a regular mesh any more. We may no longer simply imagine that the grid is displaced in space without deformation. The idea that comes to mind at this juncture is to consider the grid as a realization of a point process. If, in addition, this process is considered as stationary, we obtain once more, for the estimation variance, formulae where the REV appears only through its transitive covariogram g(h). Here again the principle of the method is to attribute stationarity not to physical reality (the REV) but to the information device (the grid).

Clearly one must closely examine the meaning of such a hypothesis, or rather, methodological decision. For this is *a constitutive decision that, in particular, specifies the estimation variance as a function of the characteristics that we choose to attribute to the point process*. In order that the concept so constructed may be considered as realistic and useful in practice, the survey grid must be sufficiently homogeneous to allow us to assimilate it plausibly to a realization of a stationary point process having such and such characteristics.

Let us note that in this case (unlike the situation where one must advance a stationarity "hypothesis" concerning physical reality) we do in general have in hand, in praxi, all the required assessment tools either if we have positioned the grid ourselves or if, at the very least, the history of the survey is available to us. There are unequivocal situations (e.g. when the positioning of the points has been actually carried out through a random draw either according to a purely random scheme or according to a stratified one). It also often happens that the grid appears to be highly heterogeneous, but can, without much ambiguity, be subdivided into two or more homogeneous sub-domains. It is then worthwhile to treat each of the subdomains separately. This situation occurs quite frequently in Mining Geostatistics, when some part of the deposit has been more thoroughly surveyed than the remainder (because it is richer, or simply for the purpose of develeopmental planning). But there are some rather difficult cases, e.g. when it is known that the positioning of the grid has been influenced by hypotheses concerning the REV (and it is immaterial whether they are correct or not). Thus, in mining practice, it is known that at the preliminary stage the drill-holes are often positioned with a structural objective in view, for example, the testing of some hypothesis of geological nature etc... This preferential effect may generate biases that are significant but difficult to assess, thus voiding any attempt at precise quantitative estimation.

From the mathematical point of view this model (namely a sampling grid considered as a stationary point process) leads to formulae that are analogous to (4). Here again one can bring to the fore the crucial role played by the behaviour of the variogram in the neighbourhood of the origin. For example, in the case of a "random stratified" mesh the following expression for the estimation variance is obtained:

$$\mathrm{Var}\, Q^* = v[g(o) - g(\Pi)]$$

where $g(\Pi)$ represents the mean value of $g(x-y)$ when the two points x and y range over the rectangle Π that defines the mesh. Only the values taken by g in a neighbourhood of the origin whose dimensions are of the same order of magnitude as the mesh size itself appear in the formula.

Changeover to the Usual Probabilistic Models

In practical applications one rarely uses the transitive representations themselves. It is preferable to replace them by probabilistic models of the usual type that are easier to apply. The reason is that transitive representations operate with integrals rather than with mean values, making their handling rather heavy. Consider, for example, the definition of the transitive covariogram in Eq. (3). The integral extends in fact over the domain S(h) where z(x) and z(x+h) are both different from zero [S(h) is the intersection of the domain S and the domain derived from it by the translation −h]. We denote by K(h) the measure of S(h). The numerical value of the variogram is very much influenced by the function K(h) which, in turn, reflects the geometric properties of the domain S rather than the variability of the REV z(x) itself. It is therefore of interest to attempt to separate these two effects by replacing the integral by the corresponding mean value. Let us therefore set

$$C_R(h) = \frac{1}{K(h)} \int_{S(h)} z(x)z(x+h)dx = \frac{g(h)}{K(h)}. \tag{7}$$

In certain cases this separation is illusory. For example, if the REV has a typical zonal shape that is characterized by a more or less continuous decrease from a central region of high grade, the geometry of the domain is too intimately tied to the shape of the spatial variation of the REV for these two factors to be distinguishable. In other cases, however, one has the impression that the domain S has so to speak punched out a regionalization that could have been continued much further. In the latter case, which is that of phenomena that exhibit some form of stationarity (in the physical sense), $C_R(h)$ represents the properties of the REV z "itself", considered independently of the geometry of its domain.

The function $C_R(h)$ looks like a covariance. In fact, if we choose the model "stationary RF", the expression for $C_R(h)$ given in (7) is precisely the "estimator" that would be used after the fact to carry out what the classical viewpoint calls "statistical inference" of the "real" (non-centered) covariance of the model. We know that in fact the spatial mean $C_R(h)$ exhausts the objective content of the concept of covariance, since no amount of additional experimental information will allow us to go further towards the ideal covariance. We could therefore try to identify these two functions, and choose simply $C_R(h)$ as the model covariance (2). But this way of looking at the problem misses an important point. The transitive covariogram, assuming we know it after the fact, may exhibit some small undulations, some singular points, a whole detailed structure that would furnish us with almost as much information as the REV itself. But this rich detailed structure is totally inaccessible in praxi. Even after the fact we may well be led to carry out a simplification and to replace the real g(h), which is far too complex, by a more accessible model. In particular, as we have seen, we may introduce a hypothesis concerning the behaviour of g(h) near h=0 (for example $\propto |h|^{\lambda}$). It is a hypothesis that has perfectly objective meaning[3], insofar as it implies relations such as (5) and (6) that have the status of physical laws, and may be verified experimentally. And one may also, of course, interpret this replacement of the real function g (or of C_R) by a simpler model as a crossover to mathematical expectation. It is then the simplified model, which has an objective meaning, that may serve as a definition of the covariance C(h).

On the other hand, and at least as long as we are only interested in global estimation, there is no reason to limit our choice just to stationary models. Indeed, the essential results yielded by our study of transitive representations do not depend in any way on some hypothesis concerning the physical stationarity of the phenomenon. They are only tied to the stationarity of the sampling grid. This suggests the following procedure: without any risk of experimental refutation, we may consider the REV z(x) as a realization of a *non-stationary* RF of order 2, Z(x), characterized by a covariance $C(x;y)=E[Z(x)z(y)]$ that now depends on both points x and y, and not only on their difference $x-y$. According to the classical viewpoint, this non-stationary covariance is not amenable to "statistical inference" since only one relization is available. *We* shall rather say that it has no objective meaning, and that it plays a purely conventional role in our model. This is of no importance, because we do not in fact need to know it. Indeed, in this model, the transitive covariogram becomes a random function of h, namely

$$G(h)=\int_{S(h)} Z(x)Z(x+h)dx\,.$$

The expectation of the random variable G(h) may be identified, not with g(h) itself, but with the simplified model $\bar{g}(h)$ that we have substituted for it, one which would have (for example) a behaviour $\propto |h|^{\lambda}$ near the origin, and whose physical meaning we have already highlighted. It is easy to see that this expectation $E[G(h)]=g(h)$ can be expressed in a simple manner in terms of the non-stationary covariance C(x;y). In fact, let C(h) denote the mean value of the covariance $C(x;x+h)$ of two points at a distance h apart, when x ranges over the domain S(h). We then obtain the simple relation

$$\bar{g}(h)=K(h)\bar{C}(h)\,.$$

This relation shows that it is not necessary to know the whole of the function C(x;y), but only the "mean covariance" $\bar{C}(h)$. Even more precisely, we only need to know the behaviour of $\bar{C}(h)$ around $h=0$ and its analytic nature. We may just as well use the mean variogram of the model, namely

$$\bar{\gamma}(h)=\bar{C}(o)-\bar{C}(h)\,.$$

There is the same relationship between that variogram $\bar{\gamma}$ and and the regional magnitude defined by (1) as between the function $g=E(G)$ of the model and the transitive covariogram g since $\bar{\gamma}$ is a simplified version of γ_R, having for example an analytic behaviour $\propto |h|^{\lambda}$ near $h=0$, while retaining the objective meaning and the rich physical content that we have previously described.

From the purely monoscopic viewpoint, the model that we are proposing to solve the problem of global estimation may be defined as follows: the generic model is the class of non-stationary RFs that have the same mean variogram $\bar{\gamma}$ in S and the specification of the model is given by $\bar{\gamma}$ itself, or even simply by the parameters that define its behaviour in the neighbourhood of the origin. In conclusion, therefore, we have identified and collected together under the same heading the parameters that specify the model, those that enable us to solve the given problem and those that possess the richest physical content.

Let us just make a concluding remark. In the model, the random variable obtained by substituting G(h) for g(h) is now associated with the estimation variance (4) of the transitive representation. The expectation $\sigma^2 = E(\text{Var } Q^*)$ of that random variable is what we may call the "theoretical estimation variance." Since E(G) is equal, by definition, to the model covariogram $\bar{g}$, this theoretical variance is calculated from $\bar{g}$ in the same way as Var Q* is calculated from g. In other words this theoretical variance is identical with the regular term that is tied to the behaviour of the covariogram near $h=0$, while the fluctuation Var $Q^* - \sigma^2$, which is a random variable of zero mean in the present model, coincides with the fluctuating term that is tied to the behaviour of g(h) in the neighbourhood of the range. The prerandom (and unpredictable in praxi) character of the fluctuating term is thus also accounted for in the few formulation.

Global Non-Stationary Models

As we have seen, the lack of (physical) stationarity of the phenomenon does not raise major difficulties as long as we limit ourselves to global estimation problems. But one must often pursue other objectives as well. It may happen that our aim is to study the non-stationarity itself and to give it a precise global representation, to be displayed by maps. For example, geophysicists may attempt to separate a "regional anomaly" from a "local anomaly", the former reflecting the deep structure of the substratum, while the latter is tied to superficial variations of purely local interest. This distinction may be difficult to make precise, but it certainly has an objective meaning. Physical intuition, which guides us here, points to an "auto-regulated" phenomenon. In the absence of perturbations, the spatial evolution of the phenomenon may be described by a function m(x) that is sufficiently regular, at our scale of operations, to be considered as deterministic, (that is, it has a sufficiently simple and precise functional representation). On account of irregularities and superficial variation the real REV z(x) deviates here and there from the equilibrium position m(x), but these deviations are never either wide or of long duration. It is as if a restoring force soon constrains z(x) to return to the neighbourhood of m(x), and the resulting picture is that of a sort of oscillation around the equilibrium position represented by m(x). This suggests the simple model

$$Z(x) = m(x) + Y(x) \tag{8}$$

where Y(x) is a stationary random function, of zero expectation and finite range (one can of course envisage more complex models). We shall say that m(x) is the *drift* and Y(x) the *residual* or *fluctuation*. In that model, the drift is the (non-constant) expectation of the RF Z(x) associated with the REV z(x). To the fluctuation [i.e. the RF Y(x) of the model] there coresponds a new REV, namely y(x)=z(x)−m(x). It is understood that y(x) also represents a physical reality, and not just some "error of nature." In our example it represents the influence of superficial structures. The RF Y(x) of the model is not just "noise." It too possesses natural characteristics (covariance, range etc...) but at a more modest scale than the drift m(x).

The first example is not overly ambiguous, because a preliminary physical model assures us that the dichotomy drift + residual does correspond to a reality (deep structure and surface variation). But this is not always the case. For example, in marine cartography, we expect that depth will increase as we move away from the coastline. This is manifestly a non-stationary phenomenon. But it would be rather difficult to define with any precision the concept of drift that is appropriate here. We can, of course, conventionally *choose* to call drift the result of some *smoothing* process or other (moving average, low-pass filtering etc...). But this is a purely instrumental definition, not a concept. The difficulty undoubtedly comes from the fact that the intuitive representation that we have in mind when we talk about a drift is essentially tied to a specific scale of operations. To take another example, let us consider a mountain. If we operate at the scale of ten or one hundred meters, the mountain will appear as a functional drift, and the local variation (small gullies, isolatedrocks etc...) as a residual. But if we now operate at the scale of ten kilometres, as when we study the whole range to which our mountain belongs, the mountain will appear as no more than a local fluctuation that is not distinguished by anything in particular from the other mountains that surround it. The model has changed. What previously appeared as a drift is now accounted for, at the scale of the whole mountain range, by a random function that may perhaps be stationary.

In fact there is always some give and take between the two terms of the dichotomy (8). We may, to a certain extent, choose to incorporate such and such a structural chracteristic of the phenomenon, at will, either in the drift or in the residual. There are therefore many possible models that are almost equivalent, some with a complex drift and a rather unstructured residual having a very short range, and some with a very simple drift (e.g. linear or quadratic) and, correspondingly, a richly structured residual with a complex covariance function, a longer range etc... In this case there is no unique objectivity criterion, and the monoscopic viewpoint is obligatory. It is through practice that we learn to choose, in each particular case, the type of model that allows us to best solve the problem at hand.

Thus the concept of drift, or tendency, corresponds to a false obviousness, and proves, on analysis, to be singularly equivocal. We oscillate between two poles. The first has a strong visual attraction: we instinctively think of the drift as a "regular" curve or surface that passes as nearly as possible to experimental points. By a "regular" surface, we usually mean either a polynomial of given degree (this viewpoint corresponds to the technique of least-squares fitting) or a surface that is subject (for example) to curvature constraints in order to impose on it some type of regularity (the viewpoint of the technique of spline fitting). These techniques are perfectly defensible and are very useful in some types of problems. There is no objection to them provided:

1) that we clearly recognize their conventional and purely instrumental character and that we do not interpret the numerical results they produce as either "physical reality" or the realization of some definite concept; and

2) that we do not mistake the results obtained in praxi when we apply these techniques to the experimental data only for the results that would have been obtained (after the fact) by applying them to the totality of the numerical values z(x) available in S.

This last point is most important. What is obtained by applying, for example, the technique of least squares to the experimental data $z(x_\alpha)$ is not by any means the best estimator of the result that would have been obtained by the same technique applied to the REV z if it were known at all points x of S. To obtain the best estimator, we must estimate the "future" least squares on the basis of the available information. For example, one may carry out a global kriging, if a global model is available, or more simply a splicing of local krigings if only a local model is available. It is then easy to see that estimating "future" least squares is exactly equivalent to an application of the least squares technique to the surface obtained by kriging the unknown values of z(x). In other words, *one must not carry out the least squares fitting on the experimental values* $z(x)$*, but on the kriged values* $z^*(x)$*,* $x \in S$.

The first point (that highlights the risk of plunging into pure metaphysics, unless one has at the outset some physical model) warns against the spontaneous belief in the existence of some sort of regular tendency or continuous background ("trend") in relation to which the deviations exhibited by the real phenomenon constitute "errors of nature" or, at best, "anomalies" having a purely local significance. In fact, it is difficult to see on what grounds one can accuse nature of making errors. [I am not talking here of the very different (and completely realistic) problem that arises when the experimental data are tainted with measurement errors.] The concept of anomaly is certainly rich and interesting, but is very difficult to study.

At any rate, before choosing an estimator, we must *define* in a precise and *operational* manner what we wish to estimate. Concretely, this definition may be an algorithm that is applicable, after the fact, to the REV z. It is always a regional magnitude that we should aim at estimating, i.e. at approximating as nearly as possible by means of a function of the available data. Suppose, for example, that the geophysicists ask us to separate the "local anomaly" from the "regional anomaly" on the basis of some discrete data obtained from a survey. We should then ask them the usual crucial question: if you knew the REV z(x) at every point x of the domain that is of interest to you – and that is all that is in principle experimentally possible to know about the phenomenon – how would you carry out the separation of the two components, the local and the regional? Is that distinction the result of a precise model that is suggested to you by the knowledge you may have about the physics of the phenomenon, and that would enable you, if you had an exhaustive experimental knowledge, to carry out a rigorous determination of the two components or at least to set up a well-defined estimation procedure? or would you just mechanically apply some filtering or least squares procedure? Whatever the case, we require that you should inform us of the precise numerical procedure – the *algorithm* – that you would use if you had perfect knowledge of the phenomenon, it being understood that you are responsible for the choice of that procedure.

The result obtained by applying the chosen algorithm (which is at present unknown since in fact z is only given at a few points) would be the *regional magnitude* that we, in turn, could attempt to estimate for you in the best possible manner on the basis of the available information and of the model that we have succeeded in specifying on that basis. But for our part we do not have any universal rule that would enable us, from the outset, to separate a "drift in itself" and a "residue in itself", "smoothness" and "roughness" and so on.

In certain cases, the concept of drift lends itself to an operational reconstruction, analogous to the one that has enabled us to define the range, and has then an unequivocal objective meaning. But most of the time the dichotomy (8) proves to be irreducibly arbitrary. However, we know that a sensible solution to a real problem cannot depend on a purely conventional choice. In other words, if the concept of drift cannot be objectively defined, this means that we shall never need it to solve any real problem. This remark leads us to look for models that are more synthetic than (8), and in which we give up the attempt to separate the "drift" from the "residual", but still retain an attenuated form of stationarity. The retained stationarity should be weak enough to be compatible with reality but sufficient to allow us to carry out classical "statistical inference". A good example of this kind of model is that of *"intrinsic random function of order k"*[4]. This is a random function Z(x) that is non-stationary, but whose "generalized increments" are. Among the linear combinations of the form $\Sigma\lambda_i Z(x_i)$, the only ones that are stationary are the *"authorized linear combinations"*, namely those that filter (annul) the polynomials of degree lower than or equal to k. The covariance function of Z is of the form

$$C(x,y)=K(x-y)+\Sigma a_1(x)f^1(y)+\Sigma a_1(y)f^1(x)\,. \tag{9}$$

In this equation the f^1 represent monomials of degree less than or equal to k and the $a_1(x)$ arbitrary (unknown) functions, which are inaccessible in praxi and perhaps even after the fact. The non-stationary part of the covariance, namely the two sums of the form $\Sigma a_1 f^1$, has therefore a purely conventional meaning. On the other hand, the stationary part $K(x-y)$ or *"generalized covariance,"* is accessible (after the fact) and can be estimated (in praxi). This is because the variances of the authorized linear combinations depend only on K and not on the non-stationary parts of the covariance. It is therefore the generalized covariance K that constitutes the specification of the generic model "intrinsic RF of order k". A necessary corollary of the above is that if we choose such a model we must restrict ourselves to manipulate only authorized linear combinations. This is a real limitation, but it does not preclude us from giving sensible solutions to well-formulated problems.

The concept of drift does not appear in the specification and the working tools authorized by the above model. It is nevertheless accounted for, but only implicitly. Indeed, to one and the same generalized covariance K (in a specific model) there corresponds a whole class of RFs, which differ from each other by polynomials with random coefficients. If Y(x) is one of these RFs, then the others are of the form

$$Z(x) = Y(x) + \Sigma A_1 f^1(x)$$

where the f^1 are monomials and the A_1 are random variables, which are not in general independent of Y(x). Any authorized linear combination, since it filters out by definition all polynomials of order k, will therefore define a unique variable

$$\Sigma \lambda_i Z(x_i) = \Sigma \lambda_i Y(x_i) \,,$$

independently of the choice of Y or Z. We may, if we wish, call drift the random polynomial $\Sigma A_1 f^1(x)$, which is automatically filtered out by the authorized linear combinations. But this "drift" is not truly defined, or even definable, since there does not exist, in general, in the class of RFs associated with the same generalized covariance K, any privileged random function (for example a stationary one) that could serve as a fixed reference.

In the particular case $k=0$ we obtain the class of RFs with stationary increments, defined up to a (random) constant and, apart from the sign, the generalized covariance is identical with the variogram, i.e. $K = -\gamma$. We have seen above that the behaviour of the variogram near the origin has an undeniable objective meaning and a rich physical content, and lends itself to an operational reconstruction. We may therefore speculate that this will also be the case for the generalized covariance K(h) when k is greater than 0. For reasons of convenience this question will be discussed in the next chapter that is devoted to local models.

Notes

[1] The word "transitive" is used here to allude to the notion of "transition". It indicates that we wish to take into account, in our model, the phenomena of discontinuity that are observed when we *cross the boundaries* of the domain S.

[2] Subject to checking that this function is of positive type. For although g and K are of positive type, their ratio may not be.

[3] In the sense of the Popperian criterion (falsifiability) and not of the overly strict criterion of decidability in terms of regionals (c.f. in Chap. 6 the paragraph devoted to the primary model).

[4] See my paper "Intrinsic random functions" Adv. in App. Prob., Dec. 1973 and P. Delfiner's exposition "Linear estimations of non-stationary spatial phenomena" in Advanced Geostatistics in the Mining Industry, ed. M. Guarascio et al, 1976, D. Reidel.

Chapter 7

Local Models

This chapter is devoted to local models, that is, to models where we have chosen to manipulate simultaneously only data originating from points that are relatively not far from each other. Our task will be more or less the same as in the previous chapter, namely to build a foundation for objectivity and to sketch an operational reconstruction. But it will be, in a sense, less arduous, because repetition in space, which, in the absence of repetition in time, provides us with our internal objectivity criterion, makes its appearance in a particularly natural way when we adopt the viewpoint of local models. Here again our procedure will be to start from a strictly objective model (i.e. a representation) that will later serve as a criterion for judging the objectivity and the value of the usual models. And here again the choice of the latter, as it can be carried out in praxi, will turn out to be dependent on an *anticipatory approximating hypothesis.*

Sliding Representations

In order to introduce our second group of strictly objective models, which we shall call *sliding representations,* in a natural way let us be guided by a simple problem, which I may add occurs frequently in practice, namely that of *local linear estimation.* Typically the situation is as follows: we have a REV z defined over a domain S. As usual, we continue z to the whole space by setting $z(x)=0$ for every x not in S. On the basis of the experimental points x_α, $\alpha=1,2,\ldots,N$, we wish to estimate the value of z(x) at every point of some domain S_0 *chosen by us.* More generally, we could consider the estimation of a moving average of the type $z_p(x)=\int z(x+y)p(dy)$ where p is a measure whose support is contained in the moving neighbourhood defined below. But in our introductory example we shall limit ourselves to the simplest case, namely that of the estimation of the point values z(x).

To carry out this estimation, we take a number of methodological decisions. (We shall discuss later whether they are right or wrong.)

i) We choose to only make use of linear estimators of the type $z^*(x)=\Sigma\lambda^\alpha z_\alpha$. Moreover, as in the case of kriging, we decide that to estimate a particular point x, we shall only use the experimental points that are nearest to the point to be estimated. The purpose of this decision is to increase the robustness of our procedure. To be precise, we choose a neighbourhood V of the origin, called a *moving neighbourhood,* and for estimating z(x) we only retain the points x that are situated in the neighbourhood V_x of x(V_x is the translate of V by x). In other words, if we denote by x_i the retained experimental points, and set $h_i = x_i - x \in V$, we limit ourselves to the class of linear estimators of the type:

$$z^*(x) = \sum_i \lambda^i z(x + h_i)\,. \tag{1}$$

ii) Moreover (and from the methodological point of view this is the more important decision) we decide that in order to set our estimators z*(x), we shall only use *algorithms that are translation-invariant.* This means that *the weights* that appear in equation (1) may depend on the vectors $h_1, h_2, \ldots$ of the moving neighbourhood V_x but *not on the point x* to be estimated. In other words the weights must be translation-invariant. If we call the figure formed by the point x to be estimated together with the points $x_i = x + h_i \in V_x$ that have been selected to form the estimator (1) *"the configuration"* then the weights λ^i must not be changed by a translation of the whole configuration.

I emphasize again that this is a *decision* on our part (perhaps judicious and perhaps not) and not a hypothesis about physical reality. This is a second example of a procedure that makes stationarity a characteristic (chosen by us) of the class of estimators we are using rather than a (vulnerable) hypothesis about reality.

In order to evaluate the efficiency of the estimator (1) for fixed weights λ^i and vectors h_i we take conventionally (although rather "naturally") the mean square of the error z*(x) − z(x) when the point x ranges over the domain S_0 to which we have chosen to limit our interest as a criterion. We therefore write, by definition

$$\|z^* - z\|^2 = \frac{1}{S_0} \int_{S_0} [z^*(x) - z(x)]^2 dx\,.$$

Replacing z*(x) by its expression (1) we see that the mean square error may be expressed in terms of the function C(h, h′) given by the formula

$$C(h, h') = \frac{1}{S_0} \int_{S_0} z(x+h) z(x+h') dx \tag{2}$$

in the form

$$\|z^* - z\|^2 = C(0,0) - 2 \sum_i \lambda^i C(0, h_i) + \sum_{i,j} \lambda^i \lambda^j C(h_i, h_j)\,.$$

If the function C(h, h′) were known (and in praxi we shall of course have to estimate it) the remainder of the procedure (choice of the optimal weights λ^i etc...) would then be exactly as for the case of kriging. Now according to definition (2) the function C(h,h′) is actually a *covariance.* This suggests the following *completely general* definition:

Given a moving neighbourhood V, a domain S_0 and a REV z, we shall say that the RF defined by

$$Z(h) = z(\underline{x} + h) \qquad (h \in V) \tag{3}$$

where $\underline{x}$ is the random point obtained by equipping S_0 with the uniform probability law is the *sliding representation of the REV z in S_0* (for the given moving neighbourhood V).

With this definition, Eq. (2) may now be rewritten in the form $C(h,h') = E[Z(h)Z(h')]$, so that $C(h,h')$ is actually the covariance function of the RF Z(h) just defined.

The *moving neighbourhood* V and the domain S_0 chosen by us are methodological parameters. According to the above definition, all the characteristics of the RF Z(h), $h \in V$, may be expressed in terms of S_0, V and regional magnitudes, to the exclusion of any other parameters. Our model is thus strictly objective, and therefore a representation, which is, I may add, completely tautological since all we have done till now is to express the REV z in a slightly different form. Let us however note the following points, which concern the mutual relationship between V, S_0, and the real domain S outside of which z is identically zero.

The only points y of space that are involved in definition (3) are points of the form $y = x + h$, where x ranges over the domain S_0 and h over the neighbourhood V chosen by us. They form a set that we can denote by $S_0 \oplus V$ and call the "dilatation [1] of S_0 by V." Two cases present a particular interest:

i) If the dilatation $S_0 \oplus V$ is entirely contained within the actual domain S of the regionalized variable (i.e. the domain outside of which z is identically zero), then the definition of the RF Z(h) as given by (3) does not involve the (fictitious) continuation of the REV z outside its natural domain. We shall say that we have an *internal sliding representation*. This is a model that particularly well suited to the study of the "internal" variability of z, that is, the variability that occurs inside the domain S without reference to its geometry.

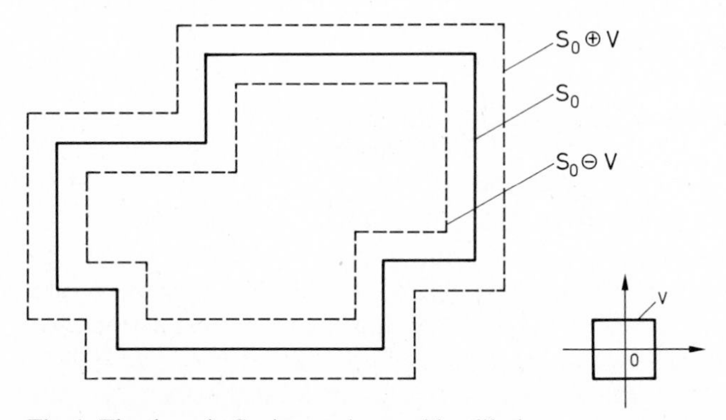

Fig. 1. The domain S_0, its erosion and its dilation

ii) Suppose now that S_0 (that is chosen by us) contains $S \oplus V$ or, equivalently, that the domain S is contained in the "erosion" $S_0 \ominus V$ (see Fig. 1). Then, since the RF Z(h) is identically zero over V every time the random point x falls outside the dilatation $S \oplus V$ of the domain S by V, all the non-trivial configurations generated by the moving neighbourhood V will effectively intervene in the definition of the RF Z(h). In that case we shall say that we have an *external* or *exhaustive representation* of the regionalized variable.

The following *theorem* can be proved: *every external sliding representation is a stationary random function over V*. In particular, if we denote by g(h) the transitive covariogram of the REV z, then the covariance function of the stationary RF over V that is associated with it in an external representation is given by

$$C(h, h') = \frac{1}{S_0} g(h - h') .$$

Nevertheless, and in spite of this pretty theorem, external representations are rarely of interest in applications. This is because they correspond to a situation where we have decided a priori to use the same translation-invariant algorithm for the internal configurations (i.e. those that are entirely contained inside the domain of the REV) as well as for those that straddle the boundary. In general this is not a realistic attitude because in most cases there is an evident benefit in treating the boundary configurations (those that contain zero values from the exterior of the real domain) separately and in a different way. Besides, it often happens that the domain S_0 that is of interest to us is a small part, chosen by us, of a much larger total domain and in that case the sliding representation is necessarily internal. In reality, external sliding representations are just a localized version of transitive representations. Just like the latter, they are only of interest when the geometry of the domain of the REV is of great importance for the problem to be solved, for example if the domain S is made up of a large number of disjoint connected components or contains numerous waste enclaves.

Priority of the Method

What the approach of sliding representations attempts to achieve is to make stationarity a (restrictive) characteristic, chosen by us, of the class of estimators to be used rather than considering it as a property of physical reality (the REV) or a characteristic of a RF model (which would not be strictly objective) chosen a priori. The attempt is completely successful in the case of external representations (see the above theorem) and only partially so otherwise. We shall return to this point later. But our methodological decision to only use estimators that have a local character and are associated with translation-invariant algorithms does not in any way presuppose either the stationarity or the homogenity of the phenomenon in space. Our decision will undoubtedly appear rather judicious if the phenomenon is actually more or less homogenous in space, but it will not be any means appear absurd if the phenomenon is heterogeneous.

Clearly, in the heterogenous case, if it is reasonably possible (on the basis of the data that are available in praxi) to divide the whole domain into subdomains over which the phenomenon remains more or less homogeneous (and to specify a different model for each of the subdomains!) then it will be worthwhile, in general, to use different algorithms, each specially fitted, for the different subdomains. The error will be reduced, on the average, and moreover the estimation variances will be *localized* (i.e. will differ from one subdomain to another, in conformity with physical intuition).

Unfortunately it is not always possible either to define in an acceptable way this partition of the domain or to specify suitable models for each subdomain. This is simply because the amount of available data becomes smaller and smaller as the number of subdomains increases. Thus the errors committed in the estimation of the objective parameters become larger and larger, and there is a substantial risk of overstepping the threshold of robustness – or even for that matter the threshold of realism.

Following our general methodological rules, we may often have to replace a powerful procedure that uses different algorithms in each subdomain, but requires numerous prerequisites, which are practically inaccessible in praxi (namely as many specific models as there are subdomains) by a less powerful procedure (a unique translation-invariant algorithm, at least for all x belonging to S), which by the same token requires fewer prerequisites. We shall of course end up with *an estimation variance that is larger and is not localized.* The numerical value that we shall attribute to it will simply represent *the mean of the values taken in the various subdomains that we no longer attempt to distinguish*. But the compensation is that we are back below the threshold of robustness and that the specification of the model is again possible in praxi.

Let us however note an important point. It is true that no hypothesis concerning the physical homogeneity of the phenomenon is required here. But on the other hand our sliding representations will be realistic only if the *survey* (the grid of experimental points) is sufficiently *homogeneous* (over the domain $S_0 \oplus V$). Indeed, for a given configuration centred at x, we calculate the estimation variance by taking the mean square of the error as x ranges over S_0. This is sensible only if we can expect to find identical configurations, or at least sufficiently analogous ones, for positionings of x that are distributed more or less uniformly over S_0. Otherwise we run the risk of making use of global characteristics of the REV that may be quite different from the local characteristics that would have been suitable for the treatment of the exceptional configuration. In case of heterogeneity the solution will be, in principle, to partition the domain into *homogeneously surveyed subdomains* (which is easier and less risky than a partition into subdomains where the phenomenon itself is homogeneous) and to treat separately each of these subdomains.

At this stage of the proceedings it is convenient to introduce a general concept, that of *sliding regional* associated with a sliding representation, of which estimator (1) is a particular case. Having continued the REV z to the whole space (by setting it equal to zero outside S), let us consider a functional $\alpha(z;V)$ defined over the restriction of z to the neighbourhood V of the origin O. For each x let us set $\alpha(z;V_x)=\alpha(z_{-x};V)$ where z_{-x} is the REV obtained from z by the translation $-x$ [i.e. $z_{-x}(y)=z(x+y)$]. We shall say that the function $x \to \alpha(z;V_x)$ thus defined is the *sliding regional* associated with the *translation-invariant algorithm* $\alpha(z;V)$. The restriction of this function to S_0 (equipped with the uniform probability law) is therefore a *random variable* associated with the RF Z(h) defined by the sliding representation, namely the random variable $\alpha(Z;V)$ obtained by replacing the REV z by the RF Z in the defining algorithm for $\alpha(z;V)$. We see that the formalism of sliding representations allows us to treat, in principle, any sliding regional, whether linear or not, and therefore, in particular, to deal with problems of *non-linear* (local) estimation, by means of conditional expectations, disjunctive estimators etc... with the sole (essential) proviso that specification in praxi be possible.

In its full generality the problem of local estimation is that of the estimation of the value over all $x \in S_0$ of a sliding regional $\alpha(z;V_x)$. If the experimental points $x_i = x + h_i$ available in the moving neighbourhood V of the point x furnish the values $z(x_i)=z_i$, then the problem is to find a function of the vectors h_i and the values z_i, or in other words an algorithm $\alpha^*(z_i;\, h_i)$, *itself translation invariant,* for the estimator. To choose such an algorithm we may adopt various criteria. The simplest is to *minimize* (in the model) the estimation variance, or, more simply, the *mean square error*

$$E[(\alpha(Z,V)-\alpha^*(Z_i,\, h_i))^2]$$

when the function α^* ranges over a family Φ *chosen by us*. Of course, in praxi, the better we can specify our model, the larger will be the family chosen. For example, Φ may be the class of linear estimators that satisfies such and such a "universality" condition, or the class of "disjunctive" estimators (see Chap. 8) etc...

The sliding regional to be estimated may be more or less arbitrary, (provided only that its definition involves only the values that are observable in the one neighbourhood V). For example, it could be a moving average

$$z_p(x)=\int z(x+y)p(dy)$$

of the REV, weighted by a measure p whose support is contained in V. The case that is most frequently encountered in practice is that of the mean grade of a "moving panel" $v_x \subset V_x$. But α may also be a nonlinear functional in z. For example, if we are interested not only in the mean grades of the moving panels but also in their distribution, we may attempt to estimate for each $x \in S$ and each real number u the quantity

$$\theta_p(x;\, u)= \begin{matrix} 1 & \text{if} & z_p(x) \geqq u\,, \\ 0 & \text{if} & z_p(x) < u\,. \end{matrix}$$

The problem to be solved, namely the estimation of the regional $\alpha(z;V)$ is given and cannot be altered. But as far as the choice of the *method* is concerned, (namely the choice of the class Φ of estimators to be used) we have a somewhat wider latitude. The larger the class Φ the more powerful will be our estimator, but also the greater the need for prerequisites relating to the spatial law of the sliding representation Z(h), for knowledge of these prerequisites is imperatively required to effectively determine the optimal estimator. If the sliding representation were completely specified (which is the case after the fact) then we should take for Φ the largest possible class, namely that of all measurable functions. It follows that the optimal estimator would then be the conditional expectation. But it is almost never possible in praxi to provide enough specification for the representation to make the conditional expectation accessible under reasonable robustness conditions. The recommended tactics are therefore to restrict the class Φ of estimators to be used in order to make the specification of the necessary prerequisites reasonably possible but without decreasing unnecessarily the power of our estimator.

We encounter again here precisely the same problem of choice and specification of a RF model as the one discussed in Chap. IV, but with one essential difference: we are now dealing with representations, that is, with strictly objective models, and the parameters of the model (except S_0 and V that have been chosen once for all) all have a purely objective meaning.

Moreover, whether Z(h) belongs to this generic model or another is itself *decidable* after the fact, so that a statement such as "the representation Z(h) is a stationary RF of order 2" is now an objective hypothesis, perhaps true and perhaps false, but decidable after the fact. Indeed, after the fact, we can evaluate the expressions

$$\begin{cases} m(h) = E[Z(h)] = \dfrac{1}{S_0} \int_{S_0} z(x+h)dx \\ C(h,h') = E[Z(h)Z(h')] = \dfrac{1}{S_0} \int_{S_0} z(x+h)z(x+h')dx \end{cases}$$

and check whether they are translation invariant or not. In practice they will never be (exactly) stationary, but we are not really that demanding, and are prepared to be content with *approximate stationarity*. If we find that, to an acceptable degree of approximation (for h and h′ belonging to V)

$$\begin{cases} m(h) \cong m \\ C(h,h') \cong \bar{C}(h-h') \end{cases} \tag{4}$$

then we need not have any qualms in replacing the real representation Z(h) by the stationary model of order 2 defined by m and $\bar{C}(h-h')$. As for what is meant by an "acceptable" degree of approximation, this clearly depends on the problem to be solved and the available data. On the basis of these data, the exact model and the approximate one must lead to estimators that differ but little, and attribute to them estimation variances that differ but little.

Obviously, it is only after the fact that we shall be able to check whether this condition is fulfilled, and whether the generic model "stationary RF of order 2" is an acceptable approximation of the representation Z(h). To adopt this model in praxi always constitutes an anticipatory hypothesis (that is both fertile and vulnerable), as in Chap. IV, but whose meaning is now purely objective (decidable after the fact) and whose epistemological status is perfectly clear: *the choice in praxi of a generic model for the sliding representation is simply an anticipatory approximating hypothesis.*

We shall now review some of the generic models that are most frequently used as approximations of sliding representations. We shall say that they are *local models*. We must not be misled by the adjective "local". Indeed, the so-called "local" models do not claim to distinguish between the various characteristics of the REV in different parts of its domain. They simply exhibit a mean picture of the various local properties. They are called "local" only because they allow us to carry out local estimations, and they are in any case only defined over the neighbourhood V of the origin. Nevertheless, all the parameters that define them have a global meaning (they are all defined by integrals over S).

We thus reach the following conclusion, which is only apparently paradoxical, and which is important from the point of view of methodology: *the specification in praxi of a local model always raises a problem of global estimation.* To handle that problem, it is usually worthwhile to go through two steps:

i) firstly one should place oneself (mentally) in the "after the fact" situation and ask what algorithm would be the best definition of a given parameter;

ii) only then can one ask how the considered algorithm (that is inaccessible in practice) may be estimated on the basis of the actually available information.

Locally Stationary Random Functions of Order 2

This is, by definition, the model constituted by the approximating hypotheses (4): namely m(h) *approximately constant* and C(h,h′) *approximately translation invariant* (in V). It is well suited to the determination of the best affine estimator of a linear sliding regional in Z (e.g. a moving average). To define m after the fact we could, for example, set

$$m = \frac{1}{V} \int_V m(h)dh\,.$$

Explicitly, m turns out to be a regional magnitude, and is the mean of the values of z(x) in $S_0 \oplus V$, weighted by a function that takes into account the distance of the point x from the boundary of the domain S_0. If the moving neighbourhood V is relatively small with respect to S_0 (and this is usually the case) then m differs but little from the (ordinary) mean value of z(x) in S_0.

Similarly the covariance $\bar{C}(u)$ may be defined as the mean value of C(h,h+u) when h ranges over the domain V(u) that is equal to the intersection of V and its translate by $-u$. Explicitly the regional $\bar{C}(u)$ turns out to be the mean value in $S_0 \oplus V$ of the expression $z(y)z(y+u)$, weighted by a function that takes into account the proximity of the boundary. Here again, if V is small compared to S_0, there will be some simplification: $\bar{C}(u)$ will differ but little from the unweighted mean of the same expression over the domain $S_0(u)$. It will just be the "mean covariance" $\bar{C}(h)=g(h)/K(h)$ of the last paragraph of Chap. 6.

It follows that the problem of the estimation of m and $\bar{C}(u)$, in praxi, will be formally the same as that of classical "statistical inference". There is nevertheless a difference that is fundamental from the point of view of methodology: we know exactly what we are attempting to estimate, namely certain space integrals, and we also know the nature of the errors that we are committing, namely those that necessarily accompany the approximate numerical evaluation of an integral on the basis of a finite number of points.

Locally Intrinsic Random Functions of Order 0

As indicated above, the class of affine estimators is usually replaced by that of authorized linear estimators of order 0, for reasons of robustness. The moment of order 1, m(h), then no longer appears among the prerequisites, and the covariance C(h, h′) is replaced by the variogram (in general not centred and not intrinsic)

$$\gamma(h, h') = \frac{1}{2S_0} \int_{S_0} [z(x+h)-z(x+h')]^2 dx .$$

The model "locally intrinsic RF (of order 0)" is then simply defined by the *approximating* hypothesis

$$\gamma(h,h') \cong \bar{\gamma}(h-h')$$

where $\bar{\gamma}$ denotes the mean variogram, whose explicit expression involves a weighted average of the $[z(x+u)-z(x)]^2$, which, if V is small compared with S_0, differs but little from the *regional variogram* γ_R defined in Chap. 6 (with the difference that the real domain S is replaced by the domain S_0 chosen by us).

Estimation in praxi is carried out in the same way as for the variogram of an intrinsic RF on the basis of a unique realization: having chosen "classes of angles and distances" Δu_i centred at u_i, we evaluate an expression of the form

$$\gamma^*(u_i) = \frac{1}{N_i} \sum \frac{[z(x_\alpha)-z(x_\beta)]^2}{2} \tag{5}$$

where the sum extends to the N_i pairs of experimental points such that $x_\alpha - x_\beta \in \Delta u_i$. One could also possibly introduce weights to take into account the "influence zones" of the various samples. It then only remains to fit to these discrete experimental values a model $\gamma(u)$ $(u \in V)$ judiciously selected. As usual, the decisive element is here *the choice of type of behaviour near the origin.*

Once divested of metaphysical interpretations the famous problem of "statistical inference" splits up into two problems:

i) *studying to what extent the discrete sum (5) is an acceptable approximation of the corresponding space integral.* This is a trivial problem of global estimation. One is, of course, never fully protected from nasty surprises, but experience indicates that in most cases, as long as the experimental points are not either too few in number or too badly distributed, the estimation can be carried out under reasonable conditions. One could moreover associate to the estimation a variance $\mathrm{Var}(\gamma^* - \gamma)$, defined within the framework of another (transitive or sliding) representation, which need not be further defined here. The evaluation of this estimation variance would require, in turn, the specification of an approximate model for the new representation, and we would be headed for another nasty round of infinite regression.

ii) choosing a type of model (that is, basically, the type of behaviour of γ near the origin) and fitting it to the experimental values $\gamma^*(u)$. The fitting itself does not raise serious problems, but that is not the case for the *choice of type:* this choice really constitutes an anticipatory hypothesis that is both vulnerable and fertile. Here again this hypothesis has the status of an *anticipatory approximation.* The situation is basically the same as for the case of the estimation of the transitive covariogram (see Chap. 6). The true $\bar{\gamma}(u)$, if we knew it, would have an extremely complicated behaviour near the origin and could not be (rigorously) simplified to one or another of the very simple types that we ordinarily use. But we could, as a first approximation, replace it by a variogram γ of simpler type, and then judge whether the approximation is acceptable by studying its effect on the problem that occupies us. If the substitution modifies the weights and the kriging variance for the configurations that are of interest to us only a little, then the approximation we made may be considered as objectively valid. In the real situation, i.e. in praxi, we advance the same approximating hypothesis in an anticipatory capacity, and, I may add, at our own risk, since here we may well be wrong, and it is only after the fact that we can verify the validity of the hypothesis.

From the epistemological point of view it should be noted that such an approximating hypothesis [namely the replacement of the real $\bar{\gamma}(u)$ by a $\gamma(u)$ of a simpler type] is not only legitimate but also, in some sense, obligatory. Indeed, if we recall the physical meaning of the magnitudes involved, we can note that beyond some scale of smallness the very notion of REV becomes fuzzy and ill defined and at that scale the concept of analytical behaviour of the real $\bar{\gamma}(u)$ ceases to be operational. But although the simplification or typification of the variogram is obligatory for physical reasons it is nevertheless certain that the type we choose in praxi is much simpler than the one we would have adopted after the fact, so that we do actually introduce an (anticipatory) approximating hypothesis.

Apart from the problem of "statistical inference", whose epistemological status we have just clarified, Mathematical Statistics also considers that of "hypothesis testing". From the very "positivist" viewpoint that we have adopted, we refuse to consider these problems at the level of generalities. We hold that there is no point in estimating a parameter that has not been previously defined in a precise operational manner, and, similarly, we refuse to test hypotheses that are devoid of objective meaning. In the case at hand the variogram to be estimated is defined in a precise manner by a spatial integral. Similarly the "intrinsic hypothesis" that we would have to test is the approximating hypothesis $\gamma(h,h') \cong \bar{\gamma}(h-h')$. It has an objective meaning since we can always determine after the fact whether this is an acceptable approximation (in relation to the solution of a given problem). But it is not easy to test it precisely in praxi through the use of well-defined criteria. This is because it is possible in general to estimate reasonably in praxi $\bar{\gamma}$ (which is a function of the one variable u), but not $\gamma(h;h')$ (which is a function of two variables). In general the available data do not allow us to obtain significantly different estimates for $\gamma(h;h^1)$ and its translate $\gamma(h+u;h'+u)$.

In fact, possible departures from stationarity or from the intrinsic character of the sliding representation are only due to the shape of the REV in the *boundary region* between the dilatation $S_0 \oplus V$ and the erosion $S_0 \ominus V$. But this region usually does not contain enough experimental points to enable us to test whether it is homogeneous or heterogeneous with any kind of precision. In practice, however, an experienced geostatistician will not hesitate much. To start with, the non-numerical (qualitative) information he possesses about the physical phenomenon indicates to him, in general, whether it is reasonable to assume that the proximity of the boundary will or will not have a significant influence. Moreover, the appearance of the "raw variograms" [the $\gamma^*(u_i)$] in different directions will usually be quite revealing. Incoherent and unstable types of anisotropy or strongly convex shapes contrasting with a more moderate rate of increase observed in other directions are sure indications of lack of stationarity. Conversely relatively stable variograms of concave or even linear shape, anisotropies of simple type etc... will encourage him to adopt the locally intrinsic model.

So far we have adopted, in the interest of simplicity, the strong criterion of objectivity (decidability instead of just the Popperian falsifiability criterion) according to which a statement is objective if it may be uniquely declared true or false after the fact, that is, when z(x) is known for all $x \in S$. But the problem of the behaviour near the origin of the "true" variogram $\bar{\gamma}$ forces us to rectify that position, and to go back to the Popperian criterion of falsifiability. Indeed, for the same given (non-intrinsic) $\gamma(h;h')$ associated with the sliding representation of a REV z, there are several possible ways of defining a mean $\bar{\gamma}$, all suitable for dealing with a given problem with a given degree of approximation. The one we have chosen is only one of these possible definitions of $\bar{\gamma}$, even though it is the simplest. In other words the model "locally intrinsic RF" is not uniquely defined and among the different $\bar{\gamma}$ that are equally compatible with reality (after the fact) there is nothing to indicate to us that one of them is "truer" than the others. It all depends on the criterion we have chosen. A given variogram model $\gamma(h)$ may certainly be found false (after the fact) if it leads to estimations that differ grossly from the ones we would obtain on the basis of the true $\gamma(h;h')$. But it can only be found true in a relative, not exclusive, sense, if it leads to estimations that differ but little. We have seen in Chap. 4 a striking example where the same set of numerical data was submitted to different practitioners and was interpreted by means of widely different models that nevertheless led to practically equivalent solutions of the given problem. One cannot say that the various models were verified, since only at most one of them could be found true, but only that they have been corroborated (not refuted). Consequently it is actually the falsifiability criterion that should be applied here.

Let us note, however, that whatever the definition of $\bar{\gamma}$, we have, in fact, replaced it by a model γ that is characterized by some type of behaviour near the origin (nugget effect, terms in $|h|^\lambda$ etc...). This concept of (analytic) type is mathematically well defined but from the physical viewpoint it must be completely reconstructed in purely operational terms. This is because the mathematical definition involves passages to the limit to which nothing corresponds in physical reality. This operational reconstruction is done in the same way as for the transitive variogram, and there is no point in repeating it.

Locally Intrinsic Random Functions of Order k

Here is a local adaptation of the non-stationary model we encountered at the end of Chap. 6. It is used when the model "locally intrinsic RF of order 0" definitely does not fit. Then we can still further restrict the class Φ of linear estimators by imposing on their coefficients "universality conditions" of a given order k. These conditions simply express the fact that *the estimators must be authorized linear combinations* of the given order (that is, they filter, or annul, polynomials of degree less than or equal to k). These estimators are *exact interpolators* for such polynomials [in the sense that they give the exact value of z(x) when the REV is itself a polynomial of degree $\leq k$].

At the end of the previous chapter we have seen that an intrinsic random function of order k should rather be considered as a class of RFs admitting covariances of the form

$$C(h,h') = K(h-h') + \Sigma a_1(h) f^1(h') + \Sigma a_1(h') f^1(h) \quad (6)$$

where K is a "generalized covariance", which characterizes the model, while the functions a_1 depend on the version chosen and remain inaccessible in praxi. The variance of an authorized linear combination, and in particular the estimation variance, only depends on K and not on the non-stationary part of the covariance.

The above remarks suggest the following definition for the model "locally intrinsic RF of order k": it consists in assuming that the covariance C(h,h′) of the sliding representation (approximately) verifies Eq. (6) for a given generalized covariance K. This is again, as is the case for all local models, an *approximating hypothesis* that is adopted in praxi in an *anticipatory* capacity and which can be checked, after the fact, by verifying that it modifies but little the solution of the problem we are interested in.

Let us note in passing that with this definition the conditions of universality that we have imposed on our estimators may no longer be interpreted as conditions of "unbiasedness". Unbiasedness is, in any case, a property that is defined only in the model and can only have an objective meaning after it has undergone the usual reconstruction. Our conditions are, in physical terms, conditions relating to filtering. Their effect is to eliminate the influence of a possible "drift", namely a component that is sufficiently regular to be assimilated *locally* (that is, over a neighbourhood V of each point x) to a polynomial of degree $\leq k$.

The model is specified by the generalized covariance K(h) alone, and it is in fact sufficient to define it over the neighbourhood 2V (double the size of V) of the origin. We have no need to know the functions $a_1(h)$ of Eq. (6). It is possible, after the fact, to evaluate them objectively, but certainly not to estimate them in praxi in any reasonable way. As far as the generalized covariance is concerned, its objective meaning is the same as that of a local variogram and its operational reconstruction may be carried out in a similar way. Its estimation in praxi (from the authorized linear combinations that can be formed with the experimental data) is very often possible under acceptable conditions.

These rather more complex questions cannot be treated here and the reader is referred to the specialized literature.

Note

[1] I am assuming that V is a symmetric neighbourhood of the origin 0 of the coordinates, so as to avoid the necessity of having to distinguish between V and its symmetric image with respect to the origin.

Chapter 8

Is Conditional Expectation Operational?

The local models that we have just reviewed are all models of order two (that is, they are specified by a covariance, a variogram or a generalized covariance) and allow us to tackle only *linear* estimation problems. The class Φ to which we have decided to restrict our search for an optimal estimator is, by the very fact that we habe adopted one of these models, the rather sparse class of authorized linear estimators of such and such an order. But practice often raises problems whose solution requires more powerful, non-linear, estimators. Let us give two examples:

i) When we are kriging, for example, a point x on the basis of the available experimental points in the moving neighbourhood V_x, we equip our estimation with a kriging variance that we know how to calculate. But, as we have already noted, this variance is a global parameter, not a local one. It is not particularly related to the structural peculiarities of the REV in the neighbourhood V_x, but represents simply the average mean square of the errors one would commit by estimating each one of the points of S by means of the same configuration of experimental points. If it happens that in the neighbourhood of the considered point x the REV behaves more erratically or more regularly than on the average, we may expect, on physical grounds, that the error will be larger or smaller at that point respectively. But since our estimation variance has only a global meaning, it cannot account for the above effect. Within the framework of a sliding representation it is impossible, by construction, to give a local meaning to the estimation variance. But even if we cannot *localize* the error (that is, express it as a function of the point x to be estimated) we may still hope to be able to *conditionalize* it (that is, to express it as a function of the values z_i observed at the experimental points $x_i \in V_x$). This conditional variance could be a satisfactory substitute for the localized variance that is inaccessible within the framework of the model. Indeed, if the REV is more or less dispersed in the neighbourhood of x than elsewhere, then the experimental values z_i observed in V_x will themselves be more or less dispersed than on average, respectively, and consequently, if the model fits, the conditional variance may be able to account for this local effect.

ii) The second example comes from a pollution problem (there would be an analogous, but more complex, formulation for the problem of selective mining).

Having measured the pollution levels z_α at some points x_α, we are seeking the probability that the mean pollution level

$$\bar{z}(v_{x_0}) = (1/v) \int_v z(x_0 + x)dx$$

in some neighbourhood v_{x_0} of a given point x_0 is higher than some alarm threshold z_0. With a sliding representation (assuming that $v \subset V$), we must therefore evaluate the probability $P[Z(v) > z_0]$ conditioned on the observations $Z_i = z_i$ available in the moving neighbourhood V.

These two problems, as well as others analogous to them, require conditional laws for their solution. As usual, before putting to use a mathematical concept, we must ask two questions: is it possible to redefine the concept in a purely operational way (after the fact) and can we estimate it in a reasonable manner on the basis of the actual information (in praxi)? The answer to the first question will be only partially positive. The answer to the second will be categorically negative as soon as there are more than one or two conditioning points, and the conclusion will be as follows: *even when we claim and think that we are using conditional expectation, that is, that we are searching for our optimal estimator in the immensely vast class of all measurable functions, what we are doing in reality is quite different, and the space Φ of functions that we are really searching is always far more restricted.* Conditional expectation, as it can be defined after the fact in operational terms, is always inaccessible in praxi (even when we harbour the illusion that we have attained it) and we always end up by replacing it (consciously or unconsciously) by expressions that are much simpler, and involve some degree of gross approximation.

After the Fact Objectivity of Conditional Laws

Consider a sliding representation, that is, the RF defined by $Z(h) = z(\underline{x} + h)$, $h \in V$, and $\underline{x}$ uniformly distributed in S. Let us investigate, within this strictly objective framework, the form of the various conditional laws that we might be interested in. Without entering into details of mathematical formalism, it is clear that for a given $h \in V$ and a given real number ζ, fixing the value $Z(h) = \zeta$ of the RF Z(h) is equivalent to forcing the generic point x to range over the set $\{x : z(x+h) = \zeta\}$ instead of the whole of S. This set is the translate $L_{-h}(\zeta)$ by $-h$ of the level curve (in R^2) or level surface (in R^3) $L(\zeta) = \{x : z(x) = \zeta\}$. Therefore, after conditioning, the generic point $\underline{x}$ is no longer uniformly distributed in S, but has a conditional distribution *concentrated on the curve* $L_{-h}(\zeta)$. If now f is a random variable of the model, that is, a function measurable on S, its conditional expectation is no longer $(1/S)\int_s f(x)dx$. It is now given by an integration over the curve $L_{-h}(\zeta)$.

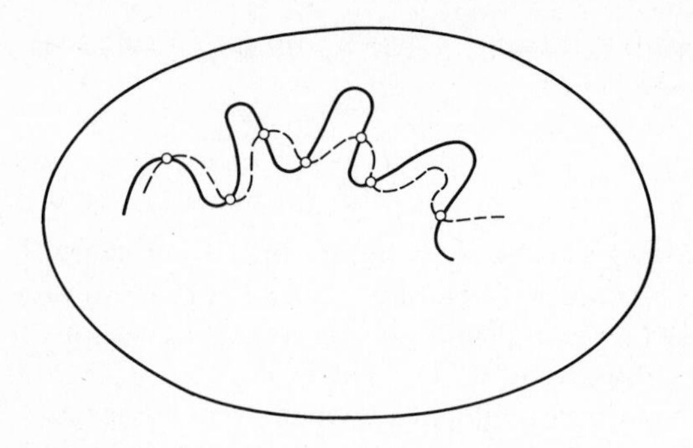

Fig. 1.

Similarly, if we condition on two points h_1 and h_2, fixing $Z(h_1)=\zeta_1$ and $Z(h_2)=\zeta_2$ is equivalent to restricting the generic point x to range over the intersection

$$L_{-h_1}(\zeta_1)\cap L_{-h_2}(\zeta_2)$$

of the translates by $-h_1$ and $-h_2$ of the two corresponding level curves. In R^2 this intersection usually consists of isolated points. Their number will be infinite if we attribute a highly pathological microbehaviour to the REV (e.g. if it is analogous to the trajectory of a Brownian motion) but it will be almost always finite if we make a physically plausible hypothesis about this microbehaviour. In R^3 the same situation will occur with three conditioning points and, in general, in R^n with n points. If we now choose p conditioning points, where p need not be large, but is greater than n (e.g. $p=9$ in R^2 or R^3) then we can expect that the intersection

$$L_{-h_1}(\zeta_1)\cap\ldots\cap L_{-h_p}(\zeta_p)$$

will be empty or will contain only one point (except perhaps for certain exceptional choices of the values $\zeta_1,\ldots,\zeta_p$). This means that there is, in general, *a single value* of x such that $z(x+h_1)=\zeta_1,\ldots,z(x+h_p)=\zeta_p$. In other words, the corresponding conditional laws are *degenerate*. Now to any given function f, there will correspond a function $F(\zeta_1,\ldots,\zeta_p)$ such that

$$f(x)=F(z(x+h_1),\ldots,z(x+h_p))$$

(except perhaps for some exceptional points x), since knowledge of the numerical values $\zeta_i=z(x+h_i)$, $i=1,\ldots,p$ suffices to determine the point x [and therefore the value f(x) of the function F at that point]. Obviously, the structure of the function F will be extremely complicated, and we will have no hope whatsoever of evaluating it in praxi.

In fact, this (degenerate) conditional law will be highly unstable, and its *physical reality* is questionable. Indeed, the point x is now a function $x(\zeta_1,\ldots,\zeta_p)$ of the conditioning values. But this function is not defined everywhere, since not all configurations $z(x+h_i)=\zeta_i$ necessarily exist. Moreover, one can expect that small variations of one or the other of the conditioning values will result in considerable variations of the position of the point x, which may for example jump from one end of the domain S to the other.

These characteristics of conditional laws that we have thus brought to the fore are closely tied with the microbehaviour of the REV z(x), and there is good reason to believe that at the extremely fine scale at which we are led to operate *we have in fact long since overstepped the threshold below which the very concept of REV ceases to be operational.* Indeed, the fine properties of level curves (and of the intersections of their translates) are neither accessible nor even really definable on an experimental basis. Just as when we critically examine continuity and differentiability, we must go back to the Popperian criterion – and apply it to the REV itself.

Within the Popperian framework some simplification occurs, but the essential conclusion, namely that conditional laws and expectations are extremely unstable, remains true, as the following analysis shows. Firstly, in order to take into account the fact that the numerical value of z(x) is only known (or even definable) experimentally with a small number of significant figures, we must *discretize* the parameter, that is, retain only a finite number of *classes* of values C. The level curves $L(\zeta)$ are then replaced by *bands* $L(C_i)=\{\chi''\zeta\{\chi\}\in C_i\}$ having a finite width. Similarly, since we cannot handle experimentally an infinite number of points, we will have to discretize the domain S itself. For example, S will have to be considered as the union of a large number of small contiguous squares or cubes, and the accessible experimental values will no longer represent point values, but rather means over these little squares or cubes. In this discretized version, we can no longer affirm that the point x is determined by the knowledge of the discretized values of the ζ_i. This will depend on the number of classes retained. An intersection of the type

$$L_{-h_1}(C_{i_1})\cap\ldots\cap L_{-h_p}(C_{i_p})$$

may, indeed, possibly contain several small squares. But then these small squares will usually be very near to each other. Sometimes, however, for certain choices of the discretized conditional values, several clusters of little squares, situated at different emplacements, will appear. Moreover, it will often be enough for one of the ζ_i to change from one class of values to the neighbouring one for the square or the cluster of squares to undergo considerable displacement in S, or even to disappear, and for new clusters to appear in unexpected emplacements etc... This extreme instability may be visualized on a texture analyser.

Thus conditional laws and expectations, even though they are definable in principle, would be far too complicated to be ever of use in practice even if we knew them, on account of their instability and extreme irregularity. Moreover it is clear that it will never be possible to estimate them in praxi. Their illusory character can be easily demonstrated by examining some orders of magnitude.

Some Orders of Magnitude

In order to shed some light on the orders of magnitude involved, let us consider the following schematic example of a trivial case of two-dimensional mine surveying. The available data (the drill-holes) represent mean grades over a support $s=20$ sq.cm. and are positioned on a 100×100m grid. The domain S of interest to us has an area of 200 hectares, i.e. $S=10^9$ s. The moving neighbourhood V is a square of 300×300m. and therefore contains 9 drill-holes. Its area is $V=4.5 \times 10^7$ s. The deposit area is discretized into little squares of 20 sq.cm. (and we assume that each drill-hole can be identified with one such square). The grade values are discretized into C classes (e.g. $C=5$, 10 or 20).

Since our moving neighbourhood contains 9 drill-holes, we are attempting to calculate expectations conditional on $Z_1,\ldots,Z_9$. But 9 variables, each taking C values, give rise to C^9 possible configurations, namely:

$$\begin{aligned} 5 \times 10^{11} &= 500 \text{ billion} \quad \text{if} \quad C=20, \\ 10^9 &= \text{one billion} \quad \text{if} \quad C=10, \\ 2 \times 10^6 &= \text{two million} \quad \text{if} \quad C=5. \end{aligned}$$

But the number of configurations actually present in the deposit is $S/s=1$ billion. It is therefore only for $C=5$ (i.e. for an extremely crude discretization) that we may hope that a significant number of the $C^9=$two million possible configurations will be represented in the deposit in numbers substantial enough to give a physical meaning to statistical reasoning. And this is after the fact only. In praxi we only have 200 configurations and no reasonable estimation is possible.

The number C^9 is also the *dimension* of the space of measurable functions of 9 variables, each taking C values. Thus with $C=10$ classes (which in may cases is already rather crude) the dimension of the working space required to determine a conditional expectation is $N=$ *one billion*. Can we seriously claim that we are really operating in such a rich space, and, in particular, imagine that it is possible in praxi to estimate one billion parameters on the basis of 200 experimental data points?

In ordinary kriging (of zero order), the dimension of the working space is 8 (6 for kriging of order 1). Between 8 and one billion there is manifestly room for some intermediate approach. In *disjunctive kriging* (see below) the class Φ of estimators is made up of functions of the form

$$f_1(z_1)+f_2(z_2)+\ldots+f_N(z_N)$$

namely a sum of N functions of one variable (N is the number of data points available in V. Here $N=9$). With C classes of values, the corresponding space has dimension CN (instead of C^N for conditional expectation), namely, in our case, with 9 data points and 10 classes, a dimension of 90. This is already high, but is considerably more reasonable than one billion. Thus disjunctive kriging does represent a valid intermediate approach.

In Praxi: Estimation of Conditional Laws

The orders of magnitude given in the previous section are decisive. It is out of the question to actually estimate in praxi a conditional law or expectation if the number of conditioning points exceeds 1 or 2. They can be calculated, but this requires the use of a model that is infinitely more specified than really permitted by the available data. Besides, the (illegitimately) overspecified models that we may use in practice all lead to relatively simple expressions for the conditional expectation, far simpler, at any rate, than those that would have been obtained after the fact by applying the operational definition, and having very little to do with them. It thus appears that by carrying out these calculations we have grossly overstepped all thresholds of realism and robustness.

There is, fortunately, another way of looking at things. If we completely give up the idea of evaluating the true conditional expectation we may then be content with putting the overspecified model, or rather its type, to a *purely heuristic use*. Indeed, the model does *suggest* to us a class Φ of algorithms that depend on only a small number of parameters (those that specify the model), each of which is an expression (in the corresponding model) for the conditional expectation. Forgetting the interpretation, we can then set ourselves the task of choosing in that class Φ (as it stands) the algorithm that suits our problem best.

This highly monoscopic way of proceeding clearly raises some questions. Firstly, since the overspecified model type only plays a heuristic role and is not attributed any objective value, what is our guarantee that another choice would not have led us to a better estimator? More generally, there is no decisive reason why we should limit ourselves to the class Φ of algorithms associated with the conditional expectations of such and such a model type. We could just as well have chosen a priori some arbitrary class Φ, as long as it depended only on a limited number of parameters and did not lead to calculations that were too complicated. This is, in fact, one way of introducing disjunctive kriging (see below). But many other choices are possible, and this opens vast horizons to research.

Having said this, we must still keep in mind what is reasonably possible in praxi. Firstly we must ask ourselves to what extent (and at the cost of which anticipatory approximating hypotheses) we can expect to carry out an approximate reconstruction of the spatial law of the sliding representation Z(h). The answer is, I think, rather clear: at the cost of an approximating hypothesis of the type "local stationarity" we may hope to reconstitute rather well the one-dimensional law of Z(h), obtain some indications about the two-dimensional law of Z(h) and Z(h′) but no more. We will usually have no hope whatsoever of extracting from the data any credible information about laws of dimension greater than two. This conclusion leads me to say a few words about estimators of disjunctive type, already mentioned above.

Disjunctive Estimators

In many cases one would like to have an estimator that is more powerful than the linear estimators studied in the previous chapters, but that does not require knowledge of parameters as numerous, as little accessible in praxi (and perhaps also as devoid of objective meaning) as conditional expectation. As an example let us return to the pollution problem stated earlier. Suppose that we have measured the level $z(x_\alpha)=z_\alpha$ of some polluting agent in samples taken at points x_α, $\alpha=1,2,\ldots,N$. Above some threshold z_0 we speak of "contamination". We wish to estimate, within some given area S_0, what may be called the "contamination percentage". Its precise definition is as follows: for every point x we set $\theta(x)=1$ if z(x) is higher than the alarm threshold z_0 and $\theta(x)=0$ otherwise. The area of the contaminated part of S_0 is given by

$$S(z_0)=\int_{S_0}\theta(x)dx$$

and the contamination percentage is defined to be the ratio $S(z_0)/S_0$ expressed as a percentage. Since $\theta(x)$ is a regionalized variable associated with z(x), the contaminated area is itself a regional magnitude and it is this regional that we wish to estimate on the basis of our scant numerical data points. Since we do not have a global stationary model that is reasonably suited to our problem we must fall back on estimating $\theta(x)$ at each point x by means of a locally stationary model and then carrying out a splicing of the local estimations so as to build up an estimator of $S(z_0)$. If the working neighbourhood V (that must nevertheless contain enough experimental points) is relatively small compared with S_0 then the anticipatory approximating hypothesis that we have advanced by choosing the above model has a reasonable chance of turning out to be acceptable after the fact, and we may even already test it in praxi to a certain extent. We are thus brought back to the problem of locally estimating a REV of type "all or nothing" (that is, taking only the two values 0 and 1). Unfortunately linear estimators are particularly ill suited to handling such REVs as is well known to any one who has gone through the experience of attempting such a problem. But on the other hand the conditional expectation of $\theta(x)$ is inaccessible and perhaps even illusory. We are thus led to the idea of using an "optimal estimator of disjunctive type" or, more briefly, "disjunctive kriging." This estimator is chosen in the class Φ of functions of the form given below (where $z_1,\ldots,z_n$ represent the n data values available in the neighbourhood V of the point x)

$$f(z_1,\ldots,z_N)=f_1(z_1)+f_2(z_2)+\ldots+f_N(z_N)\,.$$

They are functions that are expressed as a *sum of N measurable functions of one variable*. We have already made above the remark that, from the point of view of dimension, their function space represents a reasonable compromise between the truly too sparse space of linear combinations and the infinitely too rich one of N-dimensional measurable functions. Among the functions of the class Φ we choose as an estimator the one that will minimize (in our locally stationary model) the corresponding estimation variance. It turns out that this variance can be expressed in terms of the two-dimensional laws of $[Z(x),Z(x_i)]$ and $[Z(x_i),Z(x_j)]$ of the model. We have just seen that it was not unreasonable to try to estimate in praxi univariate and bivariate laws. Thus the problem is well formulated. The following table illustrates the intermediate position held by disjunctive kriging.

	Kriging	Disjunctive kriging	Conditional expectation
Optimal estimator among functions of the form	$\sum \lambda_i z_i$	$f_1(z_1)+\ldots+f_n(z_n)$	$f(7_1, \ldots z_n)$
Prerequisites	Covariance matrix	Bivariate marginal laws	(n+1)-variate marginal laws

Theory shows that within the class Φ estimators defined above the optimal estimator is uniquely determined by a system of integral equations that it is not necessary to reproduce here. Although these integral equations are of a relatively simple type, we do not know, in general, how to solve them explicitly. One could, of course, attempt to solve them numerically. But we would then have to solve a system of linear equations of rather large dimension. On the one hand, this might turn out to be expensive and on the other hand there is a risk that the system might be rather unstable, if only because of its size. Undoubtedly, we are already very near the threshold of robustness. We may even have already overstepped it, and not realized that our model has already started to function in a partially heuristic way.

Since we have already, in all likelihood, taken that fateful step, we are tempted to go a little further, and introduce a supplementary hypothesis that will enable us to simplify considerably our system of integral equations without much additional risk and yield an algorithm that is particularly economical and easy to implement. The hypothesis in question consists in choosing an "isofactorial model". This will be concisely described in the next paragraph.

Consider first the case of (n+1) dependent Gaussian random variables $Y_0, Y_1, \ldots, Y_n$, of zero expectation and unit variance, such that all the bivariate marginal laws are themselves Gaussian, and are therefore characterized by their correlation coefficient. (We could also consider the slightly more general case where the bivariate laws are of "Hermitian" type.) There then exists a sequence of polynomials H_p, called "Hermite polynomials" that have the two following remarkable properties:

i) If $p \neq m$ and if Y_i and Y_j are two arbitrary RVs from our set (j may be equal to i, and in that case the two variables are the same), then the two random variables $H_p(Y_i)$ and $H_m(Y_j)$ are *orthogonal* (that is, they have a covariance or a correlation coefficient equal to zero).

ii) Moreover, if Y is one of our variables and f an arbitrary measurable function (subject only to the condition that the second moment $E[f(Y)^2]$ be finite), then f(Y) has a series expansion, which converges in quadratic mean, of the form

$$f(Y) = \Sigma f_p H_p(Y) \qquad (1)$$

with coefficients f_p that can be calculated very easily by using the orthogonality property given in i). We shall say that the polynomials H_p are the *"factors"* of our model.

Suppose that we wish to estimate $H_p(Y_0)$, for each index p, on the basis of the n variables $Y_1,\ldots,Y_n$. It can then be shown that the optimal disjunctive estimator H_p^* is a linear combination involving only the polynomials *of the one degree* p: $H_p(Y_1),\ldots,H_p(Y_n)$, and that the coefficients of the linear combination are the solutions of a linear system of equations of rank n (identical to the usual kriging system). Moreover, the optimal estimator of the random variable $f(Y_0)$ is obtained by replacing in the expansion (1) each factor H_p by its disjunctive estimator H_p^*, so that

$$f^* = \Sigma f_p H_p^* .$$

To put it briefly, what we do in this model is therefore *to perform a separate kriging on each of the factors.* Instead of having to solve a single linear system of extraordinarily high order (theoretically infinite since we are dealing with integral equations), we only have to solve separately as many systems of rank n (a much more modest number) as there are terms in the retained part of expansion (1). In practice it is rarely necessary to keep more than about ten terms, and disjunctive kriging reduces to about ten ordinary krigings.

Let us now return to our regionalized variables. The variables Z(x), $Z(x_1) = Z_1,\ldots,Z(x_n) = Z_n$ associated, in our locally stationary model, with the point x to be estimated and the experimental values $x_1,\ldots,x_n$ are clearly not Gaussian, except in particular cases. But one can always find Gaussian RVs $Y_0, Y_1,\ldots,Y_n$ and an increasing transformation Φ, sometimes called an "anamorphosis", such that $Z(x) = \Phi(Y_0), Z_1 = \Phi(Y_1),\ldots,Z_n = \Phi(Y_n)$. Conversely, the variables Y are determined by the variables Z through the inverse transformation and each one has separately a univariate Gaussian distribution. This does not imply, in general, that each pair has a bivariate Gaussian distribution. Nevertheless, we shall assume that they do. In more precise terms we choose a model for which this property (namely that the anamorphosis leads to bivariate Gaussian distributions) is verified and it is this generic model that we call the *isofactorial model.* The remainder of the procedure is then the same as in the Gaussian case since every function of the variable Z becomes, through the transformation Φ, a function of the corresponding variable Y. Here again, disjunctive kriging reduces to about ten simple krigings.

The above additional hypothesis (namely that of "Gaussian anamorphosis of order 2"), which we have introduced for the sole purpose of reducing the problem to the simpler isofactorial model, is not devoid of objective meaning.

Indeed, one can check after the fact, and even perhaps in praxi, whether the bivariate marginal laws of the transformed variables are approximately Gaussian or not. But even if our hypothesis turns out to be approximately verified, we are still, nevertheless, certainly overstepping the threshold of robustness, and perhaps also the threshold of objectivity, if we conclude from our hypothesis that *all* the factors are really orthogonal in pairs. Rigorously speaking, the isofactorial model is already operating at the purely heuristic level.

From the monoscopic viewpoint, a heuristic model may be considered as a *reservoir of algorithms,* into which we may search for an algorithm that appears to be suitable for the solution of the problem that are attempting to solve. Indeed, it is perfectly clear that in the absence of a suggestion on the part of the theory it would never have come to our mind to use that particular algorithm. It is in that sense that we are making a *heuristic* use of the model. This is not by any means an illegitimate use, *provided* we do not have a blind confidence in our model, and we carry out after the fact, and already, if possible, in praxi, as severe experimental tests as we can. After the fact, we will have all the time we want to compare the real values with the estimates we have made, using disjunctive kriging, on the basis of the experimental points; to confront the estimated variances predicted by the model with their actual experimental values; and finally to judge the comparative performance of the estimators suggested by different heuristic models: disjunctive kriging, conditional expectation etc... In praxi, the room for control is far more restricted. Nevertheless, bivariate laws may already be studied and the hypothesis of a Gaussian anamorphosis of order 2 may be submitted to testing. Moreover we may take each experimental point, consider it as unknown, and estimate it by disjunctive kriging on the basis of its neighours. There again we have a possibility of comparing true and estimated values as well as theoretical and estimated variances etc... By applying the controls we confer on our estimators an objective meaning. The latter is, I may add, no longer necessarily exactly the same as the meaning originally attributed to them by the heuristic models that had suggested them in the first place. But we may then carry out the operational reconstruction of the physical concepts that our estimators implicitly involve.